CULTURE, PLACE, AND NATURE
Studies in Anthropology and Environment
K. Sivaramakrishnan, Series Editor

Centered in anthropology, the Culture, Place, and Nature series encompasses new interdisciplinary social science research on environmental issues, focusing on the intersection of culture, ecology, and politics in global, national, and local contexts. Contributors to the series view environmental knowledge and issues from the multiple and often conflicting perspectives of various cultural systems.

Living with Oil and Coal

RESOURCE POLITICS AND MILITARIZATION IN NORTHEAST INDIA

Dolly Kikon

UNIVERSITY OF WASHINGTON PRESS

Seattle

Living with Oil and Coal was made possible in part by a Research Support Grant from the School of Social and Political Sciences at the University of Melbourne.

UNIVERSITY OF WASHINGTON PRESS
www.washington.edu/uwpress

LIBRARY OF CONGRESS CATALOGING-IN-PUBLICATION DATA
Names: Kikon, Dolly, author.
Title: Living with oil and coal : resource politics and militarization in Northeast India / Dolly Kikon.
Description: Seattle : University of Washington Press, [2019] | Series: Culture, place, and nature | Includes bibliographical references and index. |
Identifiers: LCCN 2018049575 (print) | LCCN 2018051562 (ebook) | ISBN 9780295745022 (ebook) | ISBN 9780295745039 (hardcover : alk. paper) | ISBN 9780295743950 (pbk. : alk. paper)
Subjects: LCSH: Mineral industries—India, Northeastern. | Natural resources—Political aspects—India, Northeastern. | Natural resources—Social aspects—India, Northeastern. | Citizenship—India, Northeastern. | India, Northeastern—Politics and government. | India, Northeastern—Social conditions.
Classification: LCC HD9506.I42 (ebook) | LCC HD9506.I42 K55 2019 (print) | DDC 333.8/209541—dc23
LC record available at https://lccn.loc.gov/2018049575

COVER DESIGN: Mary Yang
COVER PHOTOGRAPH: Traders inspecting a collapsed coal mine in Nagaland. Photograph by the author.

Dedicated to Nilikesh Gogoi (1969–2007)

CONTENTS

FOREWORD

Informed by perspectives from political ecology, over the last couple of decades, anthropological studies of natural resource extraction have often converged on some conception of a frontier. This may be a political frontier, a place on the boundaries of a nation-state where extractive economies flourish and disrupt local economies and societies. It may also be a cultural frontier, where exploited land or mineral or hydrocarbon wealth exists in the traditional territories of indigenous people, ethnic minorities, or tribes/adivasis (in South Asia). And it may also be an ecological frontier, where farming gives way to agro-pastoralism, or extensive cultivation, or simply forests on hills that are logged or protected.

More recently these ready mappings of space, territory, and society have come under closer scrutiny. Resources and frontiers are now examined for the ways they are defined, articulated, and produced by human cooperation and conflict, sometimes with nature as non-human life or vibrant matter. The role of technology and markets, alongside state building and social movements, is studied more carefully to suggest that extractive economies produce more than frontiers where nature and disenfranchised minorities are jointly deprived. In such work on resource frontiers, the constitution of resources and the making of social communities that construct social ecologies of extraction, development, and militarized sociality are receiving innovative analysis. Dolly Kikon's study of the borderlands between Assam and Nagaland in Northeast India is one such creative project based on sustained long-term research in a region in which she has worked for more than a decade in many capacities.

Kikon provides a fascinating account of how boundaries and borders are made, disputed, and maintained. Commerce, security forces, territorial anxiety of village residents, and interstate conflict marked as ethnic differentiation all generate the borders and associated outcomes; the borders are crossed in various ways to undermine distinctions and also to produce local alliances and solidarities that enable the imagination of different desired futures. Such futures are not antagonistic to national belonging; in some cases they depend on the nation-state for realization. These are some of the significant insights of the fine ethnography of the ecotonal foothill zone that Dolly Kikon provides as she traverses the Assam-Nagaland borderlands.

The most significant accomplishment of the study is its resolute focus on intimate relations, most creatively investigated through the idea of *morom* or love in two central chapters. Love for landscape and love for state are included in this capacious discussion, alongside the more predictable love of family, friend, and neighbor. Friendships and labor exchange apart from, but facilitated by, love of various kinds remain central to the rest of the study as it examines weekly markets and the imagined futures of resource-rich territories where coal is mined. In particular, the chapter on state love becomes a vehicle for commenting on multiple co-constitutive sovereignties that work radially in terms of proximity or distance from borders, but also along the elevation gradient as both plains and hills might become strong centers from which love only partially reaches the foothills.

The book is organized around key nodes of interaction, from the home and the village to the weekly market and the ubiquitous border checkpoints of various agencies, including seasonal ones that coincide with festivals. Different chapters go deeper into each of these distinct spaces of sociality, apart from the mines, oil rigs, and tea plantations. This type of close-to-the-ground ethnography is all too rare for northeastern India, where it is very difficult to do sustained fieldwork in places like Nagaland. Dolly Kikon has written an original and deeply immersive study of the borderlands of Assam and Nagaland that will take its deserved place as a significant work on the important region between Southeast Asia, China, and India.

She shows, politely but firmly, that this area is not *zomia*.[1] If it is a borderland, the production of borders certainly happens through the work of states and corporations, but also in the travels, labor, and foibles of the ordinary people who live and work in this terrain. This study brings fresh and exciting direction to the growing field of borderland studies, and it establishes the importance of paying attention to affect and emotion in natural resource extraction and conflicts, rural development, and the violence of insurgency and its suppression by militarized states.

K. SIVARAMAKRISHNAN
Yale University
May 2018

ACKNOWLEDGMENTS

The research for *Living with Oil and Coal* started in 2006. During a road trip along the Pioneer Road, Nilikesh Gogoi, Sarat Phukan, and I began thinking about the foothills along Assam and Nagaland as a militarized environment, and about the dynamics of resource extraction operations that took place here. I had already embarked on the research project that for more than a decade has led me to explore the lives lived in militarized places in Assam and Nagaland where hydrocarbon exploration and extraction take place. It is my hope that this book will help readers understand the extraordinary dynamics of power, violence, and aspirations that underlie human relationships in violent geographies marked as rich resource hubs. The detailed ethnographic accounts I present here demonstrate the layered social lives in a carbon landscape. Although my book does not deal with the theme of aspiration directly, I wish to draw the reader's attention to the conversations about promises of a secure future produced in such a landscape. I hope this will allow us to connect with the structures of poverty, insecurity, and violence that drive extractive resource regimes. Conversations about a sustainable and carbon-free planet require closer critical engagement and recognition of growing poverty, insecurities, and landlessness across resource extractive zones globally.

Starting in 2001, I had worked with civil and political rights groups to document state violence and human rights cases in Northeast India as a human rights lawyer. But post 2007, when the Central Industrial Security Forces guarding oil rigs and gas gathering stations in the foothills murdered Nilikesh Gogoi and his friend Bulu Gogoi, I began to explore how the extraordinary violence and impunity of the Indian state in Northeast India included controlling and profiling both citizens and natural resources. While I was deeply passionate about this project, I am indebted to many people who worked with me to make this research possible.

My deepest gratitude goes to the people who generously helped me navigate the foothills of Assam and Nagaland. I want to acknowledge, in particular, Kunti Gogoi, Dr. Rajan Gogoi, Dipu, Noga, Toshi, Hemo Gogoi, Ashio Patton, Mr. Kithan, Nyimtsemo, Samar, Augustine, Birsa, Nelson, Apeni, Romeo Ahmed, Qutubuddin Ahmed, Yanbeni, Yashami, Tutul, Jimbu Kondha, Easter Kondha, Shamu Kondha, Dr. K. C. Talukdar, Jayanta Chutia, Benrithing, and Chumbemo. I wrote this book in Nagaland, Assam, Sweden, Australia, and California. I am grateful to friends and well-wishers who saw me through different writing stages of this project: Duncan McDuie-Ra, Haripriya Rangan, Azung, James, Mary Lotha, Nymbeni Patton, Nchumbeni Merry, Yengkhom Jilangamba, Parismita Singh, Juliette McDuie-Ra, R. K. Debbarma, Soibam Haripriya, Lachit Bordoloi, Chandan Sharma, Senti Toy, Zhanara Nauruzbayeva, Daniel Gallegos, Tania Ahmed, Kutraluk Bolton, Oded Korczyn, Yoonjung Lee, Curtis Murungi, Steven Lee, Inessa Gelfenboym, Arup Koch, Pratiksha Baxi, Nitin Sethi, Akiojam Sunita, Tongam Rina, Akum Longchari, Aheli Moitra, Arkotong Longkumer, Mhonlumo, Yirmiyan, Kekhrie, Ningreichon, Martin, Anjulika Thingnam, Vijay Nagaraj, Omio, Indira Laisram, Laurence Belcher, Inotoli Zhimomi, Nick Lenaghan, Asojiini Rachel, Augustine, Lahun, Lallian, Ankur, and Himali Dixit.

I want to acknowledge my teachers who opened the world of research and critical thinking: Sanjib Baruah, Ranabir Sammadar, Barry Sautman, Uma Chakravarti, Urvashi Butalia, Radhika Singha, Srimanjari, Shahana Bhattacharya, Prakash, Myingthunglo Murry, and Chipeni Lotha. I am also grateful to the University of Melbourne and my colleagues at the School of Social and Political Sciences for their support during the completion of this book. I have benefited from the fellowship of Monica Minnegal, Tamara Kohn, Ana Carballo, Clayton Chin, Ana Dragojlovic, Bart Klem, Bina Fernandez, Kalissa Alexeyeff, Rachael Diprose, Kate Macdonald, Elise Klein, Adrian Little, and Fiona Haines. At Stockholm University, Heidi Moksnes, Mark Graham, Johan Lindquist, Shahram Khosravi, and Helena Wulff from the Department of Social Anthropology gave me an intellectual community. I am particularly grateful to Bengt G. Karlsson, who offered valuable insights and comments to navigate the project.

In Assam, North East Social Research Centre (NESRC) gave me office space and research affiliation. Thanks to Melvil Pereira, Alphonsus, Walter Fernandes, Mala, Rita, Ashpriya, Meenal, and Jyotikona for their support and help. At Stanford University, Thomas Blom Hansen and Sharika Thiranagama were enthusiastic and inspiring interlocutors. I remain indebted to their generosity and friendship. Sylvia Yanagisako and James Ferguson offered critical feedback. At the University of Washington Press, series editor

K. Sivaramakrishnan provided his invaluable support and played a central role in shaping the project. Thank you for reading the manuscript closely and providing important suggestions. Lorri Hagman, Michael Baccam, and the production team shepherded the book to its completion. Joel Rodrigues helped me organize the references, and Sarat Phukan worked with me to produce the resource maps of the foothills. Thank you very much for your friendship and support. My gratitude to Ankur Tamuli Phukan for helping me during fieldwork.

I received fieldwork funding from the Wenner Gren Foundation (2010) and a Mellon Foundation Dissertation Fellowship from the Stanford Humanities Center (2012). I also benefited from the Riksbankens Jubileums-fond postdoctoral fellowship when I was writing this book at Stockholm University. Some materials about the Adivasi accounts of the foothills and Samar's story in chapter 1 were published in 2017 in *Contributions to Indian Sociology*. A section on documents and friendship in chapter 5 was published in 2017 in *South Asia: Journal of South Asian Studies*. A revised version of chapter 2 was originally published as a chapter in *Northeast India: A Place of Relations* in 2017.

Finally, I would like to thank Mhalo Kikon, who has offered me unconditional love and support. Ba aka Bijoy Barbora has always given me confidence to explore new ideas and take on new challenges. Moushumi, Julie, and Rosemary have showered me with joy and warmth. Mhademo, Longshibeni, Ishaanee, Kimiro, and Samantha inspire me with their kindness and love. Kimeri and Konsang have filled my life with laughter and fun. Xonzoi (Sanjay) Barbora set me on this path of research and writing two decades ago. Without his love and support, I could never have imagined embarking on this project. This book is a witness of our journey, all the way from Jorhat, Dimapur, Guwahati, and Delhi to Hong Kong, Oakland, Stockholm, and Melbourne—the geographies that brought us together.

LIVING WITH OIL AND COAL

Introduction

I T is a November evening in 2009, and a fleet of coal trucks is marooned at the foothill border between the states of Assam and Nagaland. Sitting in a jeep with a group of coal traders returning from a Naga village, I wait along with other passengers for the Assam Police to unlock the border gate and allow passage. Every day, there is heavy traffic of people and commodities such as bamboo, timber, grains, and sand through such checkpoints between the two states. Many villages here are accessible only by gravel paths crisscrossing through tea plantations, oil drilling sites, and security checkpoints between the Assamese town of Gelakey and the Naga coal mining village of Anaki Yimsen. Police and other officials from Assam regularly lock the border gates to monitor the movement of people and goods in and out of the tribal hill state of Nagaland.[1] During the peak season—the dry winter months from October to April, when coal mining activities take place—conflicts between villagers in the foothills escalate, and state surveillance activities are elevated. The vehicles stranded that evening were all caught in this predicament.

One cannot miss the heightened presence of security agencies. While the Central Industrial Security Forces are deployed to secure the extractive resource operations at the tea plantations and the oil exploration sites, the Central Reserve Police Force, Nagaland Armed Police, Assam Police, and the Indian Reserve Battalion monitor the highly volatile border. It is routine for security guards patrolling extractive industries such as tea and oil operations to undertake joint surveillance activities with law enforcement agencies. The locked gate that November evening was a routine procedure. A coal trader named Bhaiti who was traveling with me said, "This is the work of the OC [officer-in-charge] from Gelakey police station." According to Bhaiti, the confusion and delays at this particular gate were due to a system of surveillance introduced by the newly appointed OC.

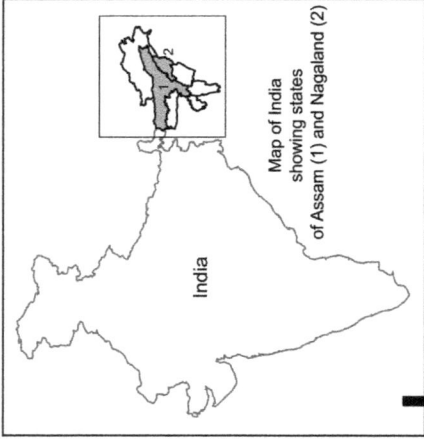

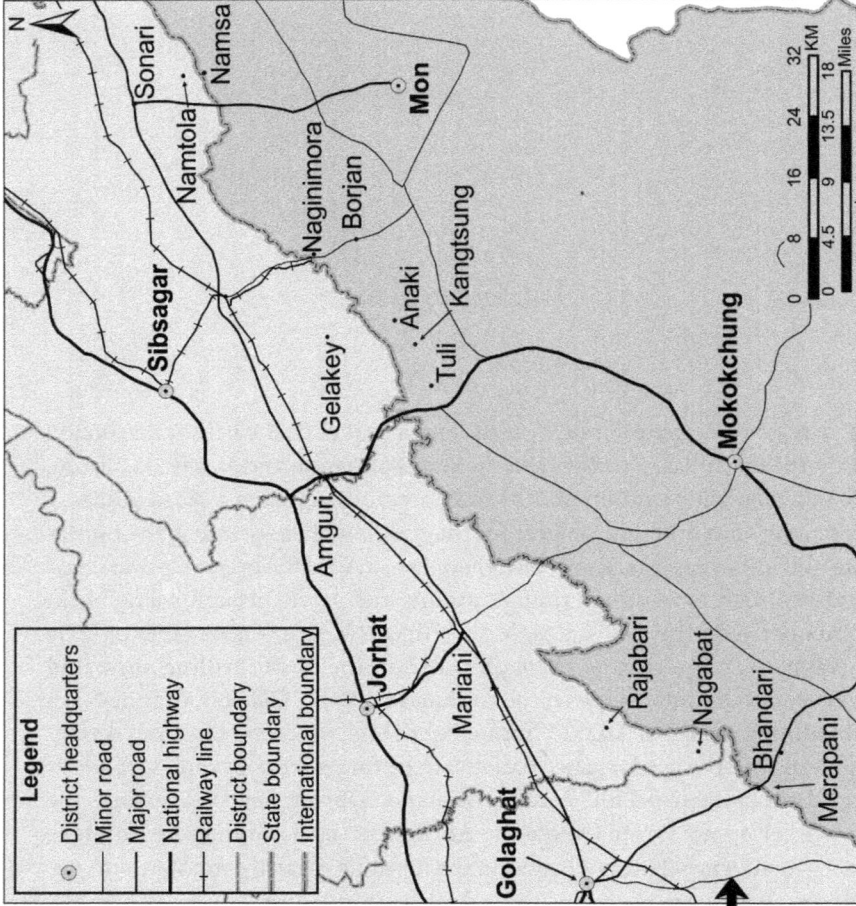

Foothill area between the states of Assam and Nagaland

The OC trusted neither his subordinates at the police station nor the plantation authorities to monitor the situation. He posted his officers at the gate but locked it and gave keys to a shopkeeper and the tea plantation watchman (*chowkidar*). The key keepers were expected to observe the police at the gate and report any bribery or inattentiveness, while the security forces were to make sure the key keepers were dutifully unlocking and locking the gate. On that particular evening, both key keepers had disappeared. Eventually, police unlocked the gate.

Surveillance and travel restrictions in these foothills go back to the nineteenth century and the establishment of tea plantations in Assam. To administer and safeguard these extractive resource operations, which eventually included coal, oil, and timber, the hills and the plains were demarcated as distinct administrative and political zones in this frontier region. British administrators imposed the Inner Line Regulation in 1873 to establish the border between the Naga villages in the hills and the plains of Assam. The Inner Line Regulation, in the words of historian Bodhisattva Kar, served as a "territorial frame to capital" (2009: 52). The regulation was set up primarily to stop British tea planters from acquiring land in the Naga Hills, which were demarcated as excluded areas and remained outside the jurisdiction of the British administration. With the exception of collecting revenue and controlling hill tribes, who were perceived as savages, the British administrators stayed away from the day-to-day functioning of the tribes and villages. The relationship between the British administrators and the hill tribes remained an uneasy one characterized by punitive expeditions, violence, and hostility.

Therefore, the colonial demarcation of the foothills was both a territorial and temporal split (Kar 2009). The division between hills and valleys was a way of designating the world of capital and modernity as belonging to the valley, in contrast to hill communities, who were perceived as tribes inhabiting a precolonial, primitive realm outside the history of development and progress. While the hills were officially left to tribal customary law, establishing them as a space of primitivism, economic development transformed the Brahmaputra Valley into a hub of global commodities such as Assam tea and oil.

After Nagaland attained statehood in 1963, the Nagas were accorded a special provision—Article 371A of the Indian Constitution—which recognized community practices, and title to natural resources. In the Brahmaputra Valley, however, with the exception of the autonomous areas, natural resources such as land, forests, and minerals belonged to the state. When I arrived in the foothills of Assam and Nagaland in 2007, I met people who

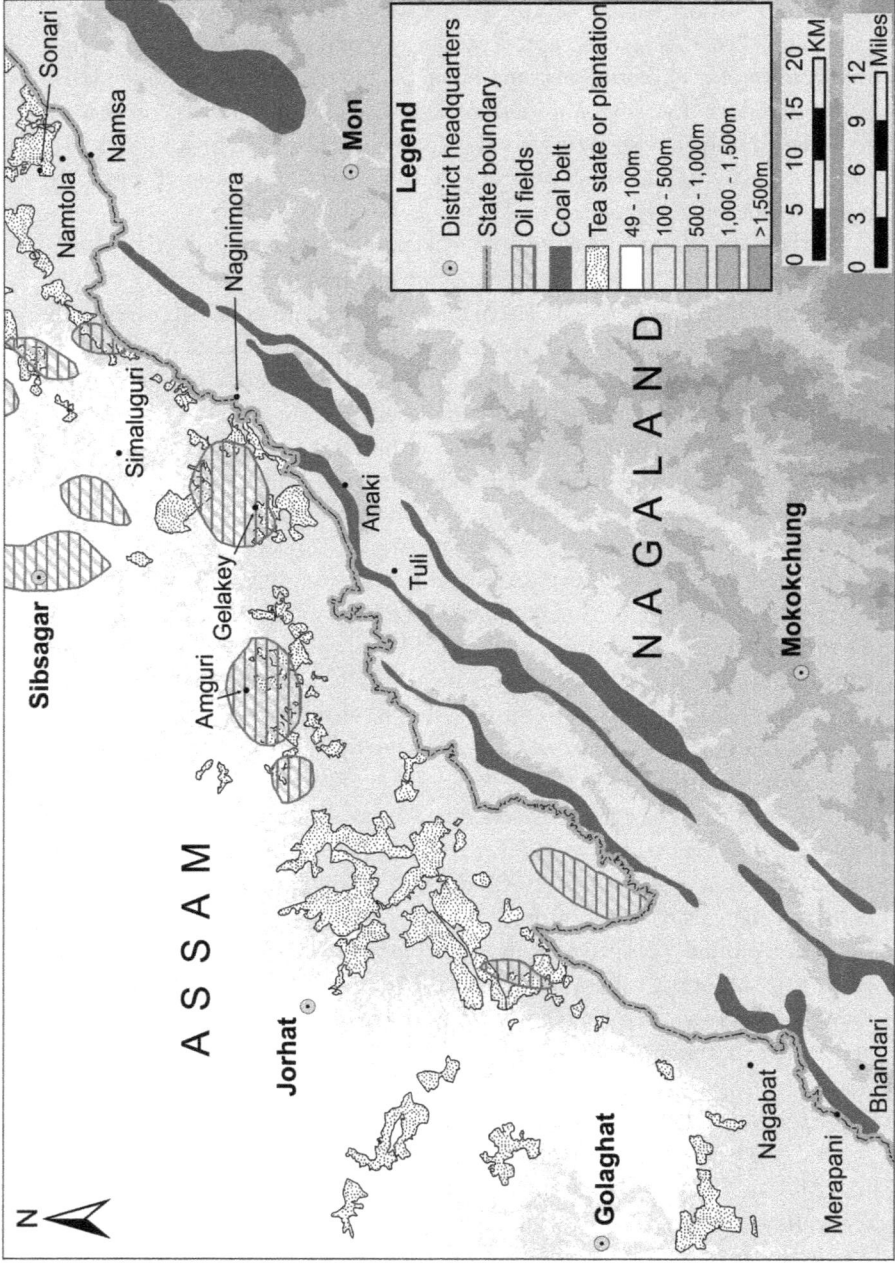

Oil wells, coal belts, and tea plantations in the foothills of Assam and Nagaland encountered various kinds of borders every day. In the villages, towns, oil drilling sites, coal mines, tea plantations, forests, rivers, and weekly markets, lived diverse ethnic groups including Naga, Ahom, Assamese, Nepali, Adivasi, Bodo, and Mishing. I traveled with them to the markets, schools, and neighboring villages, crossing rivers to get to the nearby towns. I crawled inside the coal mines, harvested rice in the *jhum* fields (fields prepared using the slash and burn method), sold vegetables in the weekly markets, watched the oil tankers pass, got frisked at the plantation and oil security checkpoints, and shared the food and household chores of my hosts in the villages.

All these activities involved crossing some kind of border. In sites ranging from a tea plantation boundary, to a barricaded oil rig, to a coal mining village, the activities and movements here constituted the political drama of resource extraction and its effect on daily life. These borders also played a significant role in producing social relations and ethnic alliances, including unequal power relations and conflicts over resources. The stories in this book trace the ways different political actors—such as coal traders, residents, state officials, and geologists—interpret the foothills as they live with the extraction of oil and coal from the earth.

CARBON CONNECTIONS

Social relations in the foothills of Assam are forged through a web of oil and coal mining activities, against the backdrop of a militarized extractive economy regime. This book, the first multi-perspective resource ethnography of an oil- and coal-producing region in India, tells the story of a place where activities and conversations continuously fuel the conflicts, alliances, and prospects of a carbon future. Oil extraction (in Assam) and coal mining (in Nagaland) produce new forms of networks, alliances, and rivalries as extractions from the ground come to the surface. Paying attention to how people's relationships with their neighbors, families, and friends—including state and nonstate actors—are produced, I avoid flattening the foothills solely as a geological and political space.

People in the foothills seldom dwell on geological maps. They told me ethnically rich and diverse stories that drastically shifted my understanding of extraction, resources, belonging, violence, and friendship. Against the backdrop of militarization and life in the Assam-Arakan oil basin (the geological name for the foothills), residents recounted their encounters with competing political actors (state and nonstate) and told stories about oil, coal, murder, death, documents, village history, love, and secrets. These insights intimately shaped the framework of this book. The geological composition of the place is often measured in scales, thrusts, folds, and belts, where different forms of petroleum—gaseous (natural gas), liquid (crude oil/condensate), semiliquid (asphalt or tar), or solid (kerogen)—were trapped. Yet many actors, including geologists, traders, cultivators, and officials, situated their knowledge of the place in different ways. From legends, songs, poems, love stories, and narratives of violence, to hydrocarbon maps, I encountered disparate images and practices around resource extraction.

The concept of living with oil and coal that is developed in this book emerged from thinking about the entangled worlds of oil and coal mining and the lives of foothill residents, but it was the work of the French philosopher Henri Lefebvre that helped ground my understanding of this concept. According to Lefebvre, although boundaries in the traditional sense represent limits and visible demarcations, in reality they are merely "an appearance of separation" (1991: 87). In actuality there is an ambiguous continuity. In the foothills dotted with resource extraction activities such as tea, oil, and coal, communities have to figure out how to navigate place despite heightened violence and militarization. Lefebvre emphasizes that boundaries enable us to recognize points of contact or ways people fashion themselves to connect with those on the other side and make sense of the world around them. In the eastern Himalayan foothills, people "interfere" with and "interpenetrate" (in Lefebvre's terms) rigid physical, social, and moral boundaries. Because community rights, including concepts of citizenship and justice, are increasingly centered on negotiations and contestations around resource extraction, these developments inform transformations in social bonds, kinship ties, and labor relations. Lefebvre asks, "If space embodies social relationships, how and why does it do so? And what relationships are they?" (1991: 27). The Himalayan foothills provide an example of how spaces of extraction are produced, and how social components and relationships come together in complex and distinctive ways.

Anthropological literature has highlighted the contestations and violence that resource extraction entails (Taussig 1980; Nash 1993; Li 2014). While my work builds on these works, which draw out the increasing political violence and transformation of indigenous societies and power relations, the stories presented here focus on internal conflicts and relations fraught with tension in communities where lives are entangled in a carbon landscape. As foothill residents, they struggle to provide for their families and have witnessed poverty, violence, and armed conflict for many decades. They see the coal mines and prospects of oil as economic solutions that will help them progress and develop their lives on this frontier of Northeast India.

In anthropology, the location of the frontier as a place of capital, extraction, and violence has become an established theme (Tsing 2005; Eilenberg 2012; Li 2014).[2] There is an especially strong focus on the contestations and crises of representation and accountability (Sawyer 2004) and scholarship around state, oil, and power (Coronil 1997; Mitchell 2011). The neoliberal state's project in Ecuador resonates with the situation in Nagaland, as both places are marked by violent struggles over land designated for hydrocarbon extraction and a history of state violence and people's protests (Sawyer 2004; Kikon and McDuie-Ra 2017). Yet, the experiences of the people in militarized places that are also regarded as natural resource hubs draw our attention to a range of hopes and yearnings for a better future. For example, several families and workers in Naga coal mining villages sought to provide a comfortable life for their children and sent them to boarding schools in neighboring towns and villages. Their motives were not limited to material gain. The villagers in the coal mining villages made convincing claims about their customary rights and resisted the state's takeover of their lands.[3]

These insights show us new imaginations and politics concerning "desired futures" that are connected to the struggles and negotiations among different actors, both state and ethnic agents, focused on the quality of livelihoods (Sturgeon 2005). The concept of desired futures is relevant to how people in the foothills of Assam dreamed of wealth and profit. But why were they attracted to a place marked by conflict and death? Nepali, Adivasi, and Bihari families struggled to find livelihood options to sustain their families. And there were also large numbers of tribal Nagas from Nagaland who sought access to markets in Assam and a place for wet rice cultivation. The high number of resource conflicts in this place, some involving violence and death, compel us to consider the lives of local people from different perspectives.

What I initially understood to be a chaotic place where administrative and political jurisdiction from Nagaland and Assam overlapped, creating routine conflicts and disasters, eventually led me to examine how residents perceived the place as a zone where different sovereign powers strictly followed procedures and orders. For example, I was drawn to examining physical encounters with representatives of the state (border magistrates of Nagaland and Assam) and figures of authority (state police from Assam and Nagaland, insurgents, and Indian security forces) as different groups navigated the resource landscape. The concept of a triadic state (Assam, Nagaland, and India; see chapter 3) emerged from my conversations with Naga landowners, Assamese traders, women cultivators, Adivasi sharecroppers, and geologists from the Oil and Natural Gas Corporation (ONGC). It is useful for illustrating how people comprehend competing sovereign powers in the foothills.

The foothills are still imagined as a place with abundant natural resources and vacant lands available for settlement, long after the discovery of oil and the establishment of the tea plantations in the nineteenth century. The prospect of the foothills as a resource-rich hub has a profound impact on life as ethnic groups, insurgents, government agencies, and corporations base their desire for homelands, new states, and markets on land and resource extraction. The experiences of the people who live there illustrate how the imagination of a place (as a resource-rich hub) and the lived reality (as a militarized zone) produce the foothills as simultaneously dangerous and desirable. These themes are evident in the following stories of land, love, friendship, and kinship, which provide an anthropological account of resource extraction in Northeast India. Claims to control of natural resources take place among contesting actors. From insurgents, state officials, tribal bodies, and private owners to mining companies, every conflict or alliance gives rise to new forms of power and transforms the politics of representation and identity in the region.

These developments enable us to redefine the social and political life of a region that has long been considered marginal in India. Rather than arguing that Northeast India is an exceptional or neglected region—a position propagated by experts, policy makers, and academics in India—this book challenges taken-for-granted categorizations such as remoteness, unruliness, backwardness, and underdevelopment. Instead, the politics of resource extraction in the foothills situates Northeast India as an important location through which we can understand new forms of heterogeneity, citizenship, indigeneity, legitimacy, and gender relations in contemporary India.

HYDROCARBON POLITICS

As everyday oil activities create alliances and frictions between the villages and the ONGC, they highlight how oil is centrally linked to the political history of resistance in contemporary Assam. "Oil exploration came to be seen as a major example of Indian internal colonialism," writes historian Arupjyoti Saikia (2011: 55) as he traces the imperial genealogy of oil in colonial Assam. The post-Independence period in Assam deeply shaped the vision of regional rights to natural resources, especially oil and gas. The Assam oil fields became a site of contentious politics asserting community and regional economic rights over the interests of the central government in regard to oil royalties in postcolonial Assam. This is a fundamental point. The question about exploitation and the economic disparity in the oil debates is founded on a larger debate about how Assam has been denied an equitable distribution of profits and oil royalties from oil production. Economist Julee Dutta argues that oil royalties highlight the unequal financial relations between the state of Assam and the central government. In Dutta's words, "Assam has been deprived of her due share of royalty on the crude produced" (2016: 49). In this logic, the contentious relationship between the state of Assam and the government of India is due to the economic exploitation of Assam's hydrocarbon deposits.

Oil and natural gas extraction in Assam has played an important role in determining the relationship between the central government of India and its citizens in Assam since India's independence in 1947. The hydrocarbon exploration that started with the discovery of petroleum in Assam around 1867 remains a high priority for India. Today, India has eighteen state-owned petroleum and gas companies,[4] including the ONGC, the largest public sector undertaking, which oversees the production of 70 percent of India's crude oil and 60 percent of its natural gas. In the foothills of Assam and Nagaland, ONGC is the most visible gas company, and its presence is an old one. In 1955, the government of India set up the Oil and Natural Gas Directorate, composed of geoscientists and experts from the Geological Survey of India. The need to establish a petroleum industry in the country led to further enhancing the power of the directorate, which was renamed the Oil and Natural Gas Commission. The mandate of the newly formed commission as per the Oil and Natural Gas Commission Act of 1959 was to oversee the development of petroleum resources, carry out geological surveys to explore for petroleum, and promote the production and sale of petroleum in post-Independence India.

Since decolonization, Assam oil fields have remained an integral part of managing natural resources, revenue, and governance in the state. According to historian Ditee Moni Baruah, in post-Independence Assam the petroleum industry became entangled with development discourse and issues of regionalism. The rise of a regional identity based on cultural autonomy and economic development, according to Baruah (2011), is intrinsically tied to control of natural resources, which involves issues of livelihood and demands for sovereignty in contemporary Assam. The issue of oil and natural gas has shaped the demands of the United Liberation Front of Assam (ULFA), an armed group fighting for a sovereign Assam. ULFA has argued that the relationship between Assam and the central government of India is predominantly economic, founded on "large-scale exploitation."[5]

The status of hydrocarbon exploration and extraction in Assam is tied to the right to self-determination, which has created what geographer Michael J. Watts (2015) calls "a permanent state of emergency." Although

Watts uses this term to describe the challenges faced by Nigeria in trying to build a multi-ethnic and multi-religious federation, he argues that hydrocarbon extraction has played a significant role in remaking power and nationalism in postcolonial societies. Referring to such political economy as "oil-fueled modernity," Watts describes the insurgencies in Nigeria as the actions of a generation that is angry and marginalized, the inheritors of resentment built over decades. The claims to natural resources in Assam since India's independence have similarly centered on oil, and people in Assam draw a parallel between state oppression and hydrocarbon extraction.

An important event that reveals Assamese anxiety over the petroleum industry centers on the movement to set up an oil refinery in Assam. The Refinery Movement (1956–57) was spurred by a proposal of the government of India to build a refinery outside the state (at Barauni in Bihar). The statewide protests stemmed from fears that the central government was taking natural resources out of Assam. This would hinder the development of Assam. People who participated in the Refinery Movement believed that a refinery would bring employment opportunities and industrial development to the state. According to geologist Sarat Phukan, these dreams remained unfulfilled despite the success of the Refinery Movement.[6] There was no development in the state from oil and gas production. Instead, increasing backwardness and poverty since the establishment of the Noonmati refinery in 1962 indicates that the state had become an enclave economy. If there was growth in any sphere, it was in the rise of technical training and skills provided in the local colleges and universities of Assam. Particularly, educational instruction at the Industrial Training Institutes helped students secure jobs in the oil and gas sector. Though not all technicians were directly employed by petroleum companies, many were employed as technical support by contractors. In addition, undergraduate and postgraduate courses in engineering, geology, chemistry, and petroleum technology helped students become scientists and professionals in oil and gas exploration and extraction operations.[7]

Nonetheless, the technical and knowledge infrastructure benefited a small number of professionals in the petroleum industry, while a larger segment of society felt growing resentment mixed with aspiration and grievance. Resentment at being excluded from the benefits of oil and gas extraction surfaced in subnational claims, where the essence of regional pride was located not only in cultural demands for autonomy but also in claims to natural resources like oil, gas, and coal (including tea). This account of hydrocarbon exploration and extraction in the state speaks directly to the task of creating a political language in Assamese civil society to advocate for managing the natural resources in the state. This is done by keeping alive the history of capital and resource extraction in Assam by attaching these connections to ongoing political conversations about India's plans to privatize the oil fields of Assam. On June 25, 2016, the Ministry of Petroleum and Natural Gas announced that it was opening up twelve oil fields in Assam to international bidders. This led to a series of protests in the state that featured famous protest slogans like "Tez dim, tel nidiu" (We will give blood but not oil).[8] During popular protests against the central government's plan to auction oil wells in Assam, activist Akhil Gogoi called this move an act of "sell[ing] out our natural resources to multinationals."[9] In 2017, the ONGC received clearance to drill ten oil wells in the foothills of Assam. All of these sites were located in Sibsagar district of Assam, and the cost of this exploratory project was estimated at 456.20 crore (approximately US$435.88 million). According to the government of India, exploration in Assam was intended to reduce the country's dependency on oil imports and increase revenue for both Assam and the central government.

The convergence of hydrocarbon extraction, protests, insurgency, and public demands for greater control over oil and natural gas reserves in the state radically shapes the vision of a carbon future in Assam: the long decades of insurgency in Assam, which have profoundly structured political and social relations among its citizens on the one hand, and the oil industry's politics of exploration and extraction, which has intimately reconfigured the militarized landscape on the other. Everyday lives in the foothills of Assam, like elsewhere in Northeast India, are regulated by Indian security forces deployed there. The extraconstitutional Armed Forces Special Powers Act (AFSPA, 1958) has radically shaped the culture of violence and impunity here (Kikon 2009). Under the AFSPA, security personnel are authorized to kill based on mere suspicion. The AFSPA spells out that "armed forces" includes both the army and paramilitary forces such as the Central Reserve Police Forces and the Central Industrial Security Forces. These security forces are posted in the foothills to protect oil rigs and gathering stations (units connected through pipelines with oil wells to collect oil and natural gas), as well as tea plantations. In addition, states like Nagaland and Assam are governed by the Disturbed Area Act of 1955, which accords extraordinary powers to the security forces to restore and maintain public order in places where the Indian state suspects public peace and safety are under threat. State agencies working in a "disturbed area" enjoy a blanket of impunity because their activities, including acts of torture and extralegal executions, are considered to have been carried out in good faith.

The militarized carbon landscape, as I discovered, was a thriving space of symbols, meanings, and people's accounts and connections with the land. In addition, the ongoing extractive activities engaged links to the past, present, and future: traders invoked the past to assert kinship ties and connections between the hills and the plains, while security forces stationed there monitored the movement of people every day, and geologists and coal miners invoked a desirable carbon future. When I arrived in the foothills, there were multiple ways to understand this place.

LOCAL LEGENDS

In the beginning of the world there were two brothers. The older one was Naga, and the younger one was Assamese. The Naga was naive, while the Assamese was shrewd. One day, their mother tied both boys to a tree and left. She used firm cane ropes to tie the older son but tied the younger one with a tender creeper. The younger brother immediately got free and caught up with his mother. In his joy, he grabbed her feet and rolled on the ground. The areas where the younger brother rolled became the plains.

After a few days, the elder brother managed to break the cane rope, but he failed to track down his mother and brother. Devastated at the loss, he hit the ground with a stick, crying and mourning. The areas where the stick struck became rivers, streams, cliffs, and mountains. A few years later, the older brother—the Naga in the hills—learned that his mother and brother were in the plains and went down to visit them. When he arrived at his younger brother's house, he saw gold, silver, salt, and abundant granaries. His younger brother was prosperous and well clothed. The opulent life of the plains attracted the elder brother, but he missed his home in the hills and decided to go back.

However, as he climbed the hills, he kept turning around to look at the plains because he was sad to leave his brother and mother. When the elder brother arrived at his home in the hills, he found that his livestock had escaped into the forest and become wild. The goat became a deer, the pig became a wild boar, and the pony became a wild horse. Even the vegetables in his garden became wild. The betel nuts in his garden became bitter nuts, and his vegetables turned into inedible plants. Since then, the inedible vegetables and plants belonged to the elder brother who lived in the hills, while the edible plants and good clothing belonged to the younger brother in the plains.

This origin myth was popular among the coal traders of Assam and in the Naga villages involved in coal mining. The retelling illustrates how local legends, colonial cartography, and administrative divisions produced the foothills as a distinctive site, but it also sheds light on a broken relationship. Attributes such as peace, harmony, and other values of sociality were often added to the myth through recollection. For example, in the Ahom and Naga villages where trading and resource extraction deals were made, this legend was told as a sign of brotherhood and camaraderie, while in locations that had experienced land conflicts, the story was narrated to reconcile the warring groups.

Another popular legend highlighted social and political relations between communities of the hills and the plains. This one centered on food. According to Pakum, the pastor of Yonlok village, many centuries ago an enemy named Maan-Singhpo came to attack the Ahom kingdom during the reign of King Gadadhor.[10] The Ahom king escaped to the hills, and the Phoms offered refuge to the king and helped him defeat the enemy.[11] After the battle, the Ahom king gifted *paan* (betel leaf) to the Phoms as a token of gratitude. This local legend also recounts that the *bhot jolokia* (Naga chili) in this part of the world came from enemies, since it was found among the food supplies abandoned on the battlefield. At first, the villagers planted the Naga chilies and wore them as ornaments around their ears until they discovered them to be edible. Today, both *paan* and Naga chilies are sold at the weekly markets known as *haats* across the foothills. Pakum concluded, "We have always been together [the hill and valley dwellers]. I say that the Assamese and the Namsang people [Nagas] can never become enemies!" Contextualizing the legend as an integral part of the political economy, Pakum said, "The *paan* leaf we eat now was a gift from the king. We continue to sustain our economy with the *paan*. It is the king's *paan*."

The Nagas have long been described as hill-dwelling people and non-Nagas as plains people from Assam. Initially, the coal traders seemed to reiterate the dominant hill and valley model. As we know, hill and valley have been important sites of knowledge production across the Himalayan region, including Northeast India. Anthropologist Edmund Leach (1954), in his seminal work on highland culture and social structures in the highlands of Burma, argues that societies are defined and reproduced by the topography and agro-ecology that sustain them. He details how study of cultivation, residence, and trade between the Kachin and their neighbors contributes to an academic understanding of the role of identity, culture, and social structure for the highland Kachin. Leach's descriptions of hill societies alternately resisting and attempting to emulate valley societies continue to affect the way relatively egalitarian, nonhierarchical societies of the hills (or what Leach calls the "Chinese model of rule of law with regard to succession and mundane daily affairs") are contrasted with their counterparts in the valley, which show more Indic influences of periodic bloodletting and violence, especially during times of transition.

Contemporary scholarship on the hill/valley continuum in Northeast India has become a rich area of inquiry. Historians, social anthropologists, and political scientists have used variations of Leach's work to highlight modern problems of governance and representation (Baruah 2005; van Schendel 2005; Scott 2009). James Scott (2009), in particular, has relied on and developed Leach's work to explain the political and historical dynamics of hill societies in upland Burma. His description of the hills as "non-state spaces" historically constituted by societies that resisted taxation and the centralized control of autocratic kings, complements Leach's emphasis on the importance of agro-ecology in understanding complex outcomes of modern state-making in places like upland Burma (and by Scott's own admission, Northeast India).

When studying people from the foothills who spoke multiple languages and created different ethnic alliances and trading networks, it became important to highlight the in-between places. The eco-tonal topography of the foothills produced new relations and practices, a point that Scott confirms as he explains how geographical features like slopes and highlands "constitute yet . . . other ecological, political and cultural zones" (Scott 1998: 187). Although Scott does not focus on the foothills, the experiences of the people I met during my fieldwork illustrate the impact of distinct geographical features.

The origin myth about the brothers is not about reiterating or rejecting a hill-and-valley framework. Instead, it forces us to reexamine accounts of escape, war, and mobility, as well as how people appropriate local legends and the past to establish social and political alliances for resource extraction in a place where the state has been present as a militaristic entity. Many people I met in the foothills were unemployed and did not have land in the hills (Nagaland) or the valley (Assam). Thus, the myth leads us to look at departure not in terms of severing relationships with certain places, but as an opportunity to reiterate a historical fracture that requires mending. Like a scene caught in a loop, residents of Naga coal mining villages and traders from Assam constantly narrated the sorrows of the elder brother's departure to the hills to reconnect and establish new kinship ties. In the same spirit, the broken bond between the brothers is used to explain misunderstandings and resource conflicts. The foothills emerge as a geographical and natural resource prism that refracts the social relations and resource extractions in the hills of Nagaland and the Brahmaputra Valley of Assam. People connect and imagine their physical surroundings in varied ways, but their attempt to gain access to natural resources produces distinct configurations of power and authority on the ground. Visible encounters take place at the numerous checkpoints located across the foothill border every day.

FOOTHILLS

Everyday life on the foothill border between Assam and Nagaland revolves around travel restrictions and curfews. Since India's independence in 1947, Northeast India has witnessed a low-intensity war. Considered the center of the longest insurgency movement in South Asia, Northeast India has been home to approximately 158 armed groups that have waged war with India since the 1950s, and a total of 36 insurgent groups are still operational, according to the South Asia Terrorism Portal.[12] In 1958, the Armed Forces Special Powers Act (AFSPA), an extraconstitutional law to aid counterinsurgency operations, was implemented in the region. The AFSPA was founded on legislation introduced by the British administration in 1942, known as the Armed Forces Special Powers Ordinance. The AFSPA suspends the protection of life and personal liberty guaranteed under Article 21 of the Indian Constitution and grants unconditional impunity to the armed forces operating under the AFSPA in disturbed areas. Section 3 of the AFSPA describes the procedures by which the government of India has the power to declare an area "disturbed" and impose the act in the following manner:

> 3. Power to Declare Areas to be Disturbed Areas—If, in relation to any State or Union territory of which the Act extends, the Governor of that State or the Administrator of that Union Territory or the Central Government, in either case, if of the opinion that the whole or any part of such State or Union Territory, as the case may be, is in such a disturbed or dangerous condition that the use of armed forces in aid of the civil powers is necessary, the Governor of that State or the Administrator of that Union territory or the Central Government, as the case may be, may by notification in the Official Gazette, declare the whole or such part of such State or Union territory to be a disturbed area.

Section 4A of the AFSPA empowers the Indian armed forces to open fire on the public to the extent of causing death, to carry out arrests, or to conduct searches on any premise without warrant if the armed forces are of "the opinion that it is necessary to do so for the maintenance of Public order."

Abandoned oil valve in the foothills of Nagaland

This legislation characterizes Northeast India as an unruly place and its inhabitants as dangerous, to be governed by the orderly rule of technocrats, scientists, and security forces.

Given this legal and political history, the foothill landscape is dotted with security agencies guarding the many extractive operations, particularly oil and tea. Security checkpoints and police stations adjacent to oil rigs and tea plantations are visible across the uneven 455-kilometer border. While Central Industrial Security Forces camps and barricades cordon off the oil rigs and tea plantations, the Central Reserve Police Force controls the Disputed Area Belt, a 10-kilometer stretch on the Assam and Nagaland border. In addition, the Assam Police and the Nagaland Armed Police are stationed at the checkpoints of their respective states. The presence of state structures representing the regional states of Assam, Nagaland, and the central government of India is overwhelming here.

During my fieldwork, I traveled from Sonari (in Assam) to Tizit (in Nagaland), foothill towns approximately ten miles apart, noting the security between these two neighboring towns. First, I came to an Assam Police checkpoint, then we were stopped at a Central Reserve Police Force post. The third gate was administered by the Assam Forest Department. Then came another Assam Police checkpoint, where the policemen checked vehicle registration documents and asked for our identification cards and addresses. A few miles later, I arrived at the Nagaland Armed Police gate, attached to the Tizit police station (in Mon district). Next, I crossed a Nagaland Forest Department checkpoint.

One's experience of the securitized foothills depended on the mode of travel. On foot, I could easily cut through paddy fields, forests, and tea plantations, where the system of surveillance and actors was more localized. Village councils erected toll gates, as did students' unions and insurgents. During the harvest festival season in the winter months (October–February), cultural associations instituted revenue collection schemes and installed temporary checkpoints. In some instances, the entry to a single village was dotted with half a dozen checkpoints, each group remarkably well organized and collecting tolls from the travelers. But these local village gates were nothing compared to the Indian security posts outside the oil rigs and the tea plantations. Armed soldiers with guns patrolled oil rigs and tea plantations; whenever ONGC geologists found oil or a tea estate cleared land to plant more tea bushes, new security checkpoints and surveillance systems appeared.

Early in my fieldwork, I saw a peculiar sight that revealed how foothill residents lived in an extractive landscape. One day, I came across a Naga man standing in the middle of a tea plantation with a crate of eggs. Soon, a woman appeared with a basket of chicken, followed by a stream of people walking through the tea plantation carrying bags of salt, vegetables, and groceries. I realized that they were returning from the *haat*, the weekly market in Gelakey. For many Naga villagers, crossing the tea plantation is the only way to come to the *haat* or visit friends in Assam. The *haats* are important sites that sustain social and economic ties in the foothills, and they are connected with origin myths and local legends. Today, many communities regard them historical and cultural signposts.

Yet many *haats* are located in precarious sites where resource conflicts regularly occur. For instance, many *haats* inside tea plantations are on land where the governments of Nagaland and Assam have overlapping claims. Some of the small *haats* are next to oil and natural gas gathering stations, while others are located next to paddy fields and rubber plantations that are demarcated as areas within the jurisdiction of Nagaland or Assam. During conflicts these *haats* shut down, and trading stops for several months. At times warring parties reconcile, but there are also cases where the matter remains unresolved or the alleged perpetrators are penalized. The punishment is a collective one where villages, clan members, or sometimes entire tribes are banned from entering the *haats* to trade for specific periods of time.

During one visit to a *haat*, I was stopped at a police checkpoint. I got out of the vehicle and asked a policeman if I could take some pictures. He nodded and signaled to go ahead. Within a few seconds, his colleagues shouted at him and ordered me to immediately put away my camera, citing security reasons. "What kinds of security reasons?" I asked. Hearing the exchanges, the officer-in-charge (OC) at the checkpoint appeared and said that it was due to the "sensitive nature of the Assam-Nagaland border issue." He was polite but appeared to be in a hurry to get back inside. There were a number of coal trucks at the checkpoint for verification of documents for the Inner Line Permit and coal papers. The Nagaland Armed Police went around looking into the windows of the vehicles, scanning both the goods and the faces of the passengers. The most visible group was a contingent of traders who frequently crossed the foothill border gates with sacks of grains, clothes, and household goods. The policemen appeared relaxed and friendly toward them. When I inquired, "What about the traders?" the OC commented that they knew the vendors as well as the owners of the coal trucks.

These security practices told a story of connections, control, and conflict. The most violent conflicts involved resource extraction and ownership of natural resources. Contests over land, rivers, and forests created a powerful structure of ethnic alliances and boundaries. While the security checkpoints and extractive operations such as oil and tea created borders here, the *real* boundaries were not drawn on the land but rather mapped on the bodies of ethnic groups who lived here and around. For example, fights between individuals turned overnight into a border fight between Assam and Nagaland. As a consequence, the essence of Naganess as a hill identity or Assameseness as belonging in the valley was magnified and reinforced through claims to natural resources. Therefore, when Nagas establish a village in the foothills, they request that the Nagaland state officially recognize their settlement as part of the hill state. Similarly, when entrepreneurs from Assam establish tea plantations, they approach the government of Assam to grant them official land documents and provide security. Timber, tea, sand, and even bamboo grooves became sites of dispute between Assamese and Naga villages. However, this was not the case with coal and oil mining areas. All coal villages were part of Nagaland, while oil and gas drilling operations fell within the territorial jurisdiction of Assam. The carbon boundary appeared immutable, unlike natural resources such as tea, rubber, or even sand, which ignited routine conflicts. Conversations about resources always came down to oil and coal, and a large part of what made the foothills attractive was its image as a hydrocarbon hub and the excitement of exploring this wealth.

THE GEOLOGY OF DESIRE

The foothills of Assam and Nagaland are part of the Assam-Arakan Basin. Geologically defined as part of the eastern Himalayan foothills, this oil basin in Northeast India is recognized as one of the oldest hydrocarbon basins in the world. Along these foothills, ONGC engineers and geologists monitor the numerous oil exploration sites and gas gathering stations. The ONGC started as a commission set up by the government of India in 1956, and became a public sector undertaking in 1994. Today, it is the largest public sector oil and gas company in India and is administered by the Ministry of Petroleum and Natural Gas. Since India's independence, extractive resource activities and the history of hydrocarbon operations in Northeast India have become targets of insurgents and ethnic nationalists because natural resources are closely tied to the right to self-determination in the region. In 1994, tribal bodies in Nagaland banned ONGC hydrocarbon exploration on their lands over conflicts around oil and gas royalties. The Naga insurgents supported the ban reiterating the right to self-determination of the Naga areas, including natural resources. Yet, the history of oil in this region is old.

Tracing the origins of geological and hydrocarbon exploration to the last quarter of the nineteenth century, historian Arupjyoti Saikia (2011) describes how the search for minerals in this frontier region of British India started with gold. In the Brahmaputra Valley of Assam, exploration for oil in Makum as early as 1837, as well as along the plains close to the Naga Hills, eventually led to the production, refinement, and distribution of oil. However, it was in post-Independence India that oil exploration and extraction in Assam became more organized and also deeply politicized. Describing the prevailing situation, Saikia notes, "At the close of the previous century the Indian government had strongly asserted that minerals and their development play a significant role in its programme of national development. In doing this, the role of the State and technology has been clearly outlined. In Assam, the oilfield townships have come to symbolise the prosperity of the Assamese elite . . . [and] oil has emerged as one of the key 'regional' resources with claims of nationalistic rights" (2011: 55). The visibility of oil townships invoked sentiments about internal colonialism. Resonating with Saikia's reflections about extraction, Assamese nationalists in the Brahmaputra Valley argue that Assam is an enclave economy of India.

Coal mining in Nagaland is relatively new compared to the long history of oil and coal activities in Assam. While small-scale coal mining operations existed in the Naga villages, new coal mines and business deals escalated after the 1997 cease-fire agreement between the government of India and the National Socialist Council of Nagaland (Isak-Muivah), an armed group fighting for a Naga homeland. During my visits to the coal mines in Naga villages, the coal mining operations, including taxation and royalties, were controlled by families connected to Naga armed groups. By the late 2000s, the debates over coal and oil operations in Nagaland touched upon issues of customary law and environmental pollution, but the most important issue related to representation. Who represented the Naga people? Was it the Naga armed group that had been involved in negotiations with the government of India since 1997 or was it the government of Nagaland? These political debates about accountability and conflict emerged from coal mining activities and dialogues about prospective oil exploration in the state.

Unlike Assam, Nagaland enjoys special constitutional provisions under Article 371A of the Indian Constitution, which guarantees that Nagas have the right to preserve their customs and culture, including the right to retain ownership of land and natural resources. Coal mining activity and discussions about resuming oil and gas exploration in the state have led to vibrant debates about land ownership, customary laws, economic change, and livelihood. This has raised important questions about governance and the changing value of land, and has led to debates between the state and the tribal bodies in Nagaland regarding communal ownership. These issues move us toward new questions about how resource control and extraction are reconfiguring conflict, political authority, and the future of politics in the region. Who owns the coal mines? What inequalities have been produced by coal, oil, and gas exploitation? Which authorities interpreting Naga customary law have allowed the emergence of a dominant nexus of landowners and political actors (state and nonstate)? In what ways are the meanings of land ownership and natural resources, as enshrined in Article 371A of the Indian Constitution, becoming a site of contestation among different authorities—armed groups, traditional councils, and the state—as to who represents the Naga people?

The ONGC report *Assam and ONGC: Synergy of Over 50 Years* gives 1867 as the date for the discovery of petroleum and coal in this frontier region and notes that the first ONGC oil well in Assam was drilled in 1959. The ONGC embraces the colonial legacy of hydrocarbon in Northeast India and knits it to the postcolonial geological and geophysical exploration of petroleum as a seamless history without a rupture. Since then, at regular intervals,

Coal mining activity in a Naga village

ONGC explorations have led to a series of major hydrocarbon discoveries in the region. For example, in 2007, of the forty-eight oil and gas deposits that ONGC discovered across India, seventeen were located in Assam (Oil and Natural Gas Corporation 2009: 7).

The Directorate General of Hydrocarbons, an official body administered by the Ministry of Petroleum and Natural Gas, describes the geology of the Assam-Arakan Basin as part of the eastern Himalayas. The hydrocarbon extraction zone in the foothills covers an area of 4,517 square miles. The production area for oil and gas in the Assam-Arakan Basin currently covers 323 miles and is divided into three geological blocks: the North Assam Shelf, the South Assam Shelf, and the Assam-Arakan Fold Belt (Oil and Natural Gas Corporation 2009). These three blocks encompass vast foothill terrains that function as the official border of Assam and the neighboring hill states of Nagaland, Arunachal Pradesh, Meghalaya, and Mizoram. Geologists searching for hydrocarbon refer to these foothills as part of the shelf-slope-basinal system. The shelf of the basin spreads over the Brahmaputra Valley, including the Dhansari Valley, while the slope of the basin stretches along the Naga foothills and the Mikir hills. The geological history and the hydrocarbon potential of this area are described as follows:

All the oil and gas fields, discovered till date in the Upper Assam shelf, are situated mostly on the southeastern slope of the Brahmaputra arch, and almost all the major oil fields like Nahorkatiya, Lakwa, Lakhmani, Geleki, Dikom, Kathaloni etc. lie in a belt bordering the *Naga thrust*. In the Dhansari valley also, oil fields like the Borholla and Khoraghat and Nambar lie in the same belt. In the *Naga* Schuppen belt, oil accumulations in the Lakshmijan and the Champang oil fields occur in that zone of the shelf, which is overridden by the Naga thrust. In the Digboi and Kharsang oil fields, oil occurs in Tipam Sandstone and Girujan Clay formations, respectively, overlying the Naga thrust.[13]

The oil fields, including the geological thrust and shelf, have been given ethnic names and meanings. For instance, the Naga thrust is named after the Naga people but has no reference to the people who live here. It is a geological term defined as "a narrow elongated zone of imbricate thrusts about 20–35 km wide extending for about 200 kilometers in the NE-SW direction. It constitutes the outmost morpho-tectonic unit of the Assam-Arakan fold belt formed as a result of the subduction of the Indian Plate beneath the Burma Plate" (Chakrabarti et al. 2011: 1). Although the sand, rock, liquid, and minerals are given ethnic names or names of people and places, the hydrocarbon exploration obliterates the social identities and political lives of the people who live there. The carbon resources, once extracted from the earth, make clear that it is not the people but the resources which are firmly connected with the sovereign project of the Indian state.

Above the ground, geologists simply refer to the villages and towns along the foothills as oil and gas fields or as zones of interest that expand and contract depending on the discovery of oil and gas deposits. Geologists working for the ONGC referred to the foothills as a range of thrusts and folds that trapped oil, but they hesitated to engage in conversations about their work. They were particularly cautious regarding topics considered political or sensitive. This reaction captured the contradictions of Northeast India as a dangerous, underdeveloped, and peripheral region but a desirable geography for resource extraction and exploration. In 1957, the same logic of political instability motivated the government of India to set up the first refinery outside the region to process crude oil from Assam in post-Independence India. Amid protests and resistance from members of the Assam Legislative Assembly and a popular uprising that demanded the plant be located within the state, a second refinery was planned for Assam. Unlike oil and gas operations in the foothills, which were perceived as being outside social and political realities, coal mines in the Naga villages attracted political and social alliances as well as the invocation of local legends. The political dynamics of coal extraction and oil exploration defined the forms and meanings of territorial power and authority. Yet these extractive activities, through which people construct their political subjectivities and alliances, significantly captured the relationship between resource extraction and militarized citizenship.

In many ways, oil exploration in Assam and coal mining in Nagaland were distinct not only in their operations and scale, but even in terms of the spaces inhabited by the respective social units. For instance, ONGC employees lived in the well-organized oil township of Nazira near the Assam-Nagaland foothills, in contrast to locals who lived in villages and towns interspersed with *haats* or local markets. The oil world appeared detached (with its fortified walls and security guards) and tranquil in comparison to the everyday conflicts defined as ethnic, local, or border problems. When their worlds collided, it was far from mundane.

MILITARIZED ENCOUNTER

I was staying in Gelakey town, an important Assamese hub where Ahom traders managed coal mining operations in Nagaland, during my fieldwork in 2009. One night, around ten o'clock, there was a commotion on the street, and the entire household came out to find a convoy of security trucks and a long line of ONGC trucks loaded with machine parts such as pipes and chains. The gigantic trucks were stuck in a residential complex and had woken up the entire neighborhood. The convoy was on its way to set up a new oil rig near Gelakey town, but the narrow streets proved challenging for the huge trucks. Such events would become common during my fieldwork in Gelakey town and the neighboring villages. Often, I heard oil trucks roar across the town, and the sound of trucks became part of my life. Yet getting used to it was not the problem, as the incident that particular night revealed. The combination of contempt, fear, and misery in the Gelakey crowd illuminated the almost unfathomably violent stories of the extractive resource regime in the foothills.

I noticed how children, parents, and youth, as well as stray dogs and cats, watched the monstrous trucks and the armed security guards. They fixed their gaze on the metal oil pipes entangled in telephone wires, television cables, and a few electrical wires. My host commented that the narrow street outside her house had become noisy and dusty since the ONGC discovered oil a few miles away. The increasing flow of vehicles carrying equipment, workers, experts, and security guards often created traffic jams. That night, the truck drivers kept their engines on, as the technicians quickly climbed up to disentangle the wires from the machines.

I was staying at Kunti's house. Her late husband, Nilikesh Gogoi, had been my first contact in Gelakey in 2006, and I had visited the coal mining villages and foothill villages with him. My first contacts with Naga coal traders and business agents and companies from Assam took place with his help in villages and towns around Gelakey. Nilikesh was killed on January 23, 2007, while returning from a coal mining village in Nagaland, before I officially started fieldwork in 2009. His violent death and the traumatic experiences of his family were constant reminders of the dangers people encountered every day in the foothills. The Central Industrial Security Forces guarding the ONGC oil and gas stations outside Gelakey town had been held responsible. Another person was shot and died on the spot with Nilikesh, while a third survived. Amnesty International condemned the attack, issuing the following statement: "Amnesty International is concerned at reports of shootings, allegedly carried out by Central Industrial Security Force (CISF) personnel, that resulted in the death of two local businessmen near Geleki town, Sivasagar district (a town outside CISF jurisdiction in Upper Assam) on 23 January 2007. A third businessman was seriously injured. Amnesty International would like to remind the Indian authorities that they are obliged, under both international human rights law and the Indian Constitution, to ensure that no one is arbitrarily deprived of their right to life."[14]

For the state agencies, these causalities became mere statistics, but residents who traversed the foothills embodied the entwinement of personal and political worlds. The following obituary described the world of Nilikesh Gogoi:

> His universe stretched from Sibsagar town to the villages of
> Anakhi Imsen—not a very large tract of land, but stable enough
> to be a storehouse of history, myths and folklore. He crisscrossed
> the winding Pioneer Road, whizzed across the Lahdoigarh Line,
> and stumbled around as though borders made not an ounce of
> difference. . . . The government and security agencies would have
> us believe that this is because there are Naga and Assamese reb-
> els in the area. Even if that were true, the government, not to
> be outdone, has thrown in its companies of army and paramili-
> tary personnel, thereby making the area a veritable garrison.
> Nilikesh saw these security forces as temporary trespassers, like
> the British planters.[15]

Nilikesh's obituary captured the everyday dangers of living in the foothills. The obituary, as much as it was a eulogy of Nilikesh's life and spirit, also remarked on the militarized structures built into the extractive resource regime. As Kunti and I stood outside her gate watching the technicians, she turned to me and said, "These oil security guards killed my husband." Her remark seemed to reflect the everyday chronicles of violence and their painful memories.

The ONGC convoy's activity unfolded like a well-rehearsed performance. Shadowy figures with guns amplified our anxieties. A number of CISF personnel, their faces covered with black masks, pointed their guns at the crowd, making it obvious that it was the trucks, machines, and technicians who required security. Kunti's neighbors, who were also watching the action, complained about the nuisance the trucks had created in the locality. The oil crew and security guards talked among themselves and kept a safe distance from the crowd as they worked to untangle the wires from the oil pipes.

Eventually, the technicians succeeded in ripping off the wires, which drooped from metal posts and swayed in a cloud of engine fumes as the trucks pulled away. We counted fifteen trucks including the jeeps carrying the security guards. Then, a white SUV nestled at the center of the convoy caught our attention. Three, well-dressed male officials sat inside the SUV, expressionlessly staring at the bystanders. Two CISF guards sat with them, pointing their guns at the crowd and shifting their gaze between the people and the drilling machines rolling in front of them. Contestations over issues of citizenship, sovereignty, and ethnic alliances in a militarized land were tied to extraction and flows of natural resources, which defined relations and shaped the outline of a carbon future.

CHAPTER ONE

Storytellers

O VERLAPPING claims by various ethnic and state actors compel residents to present the story of their village as history that is worthier than that of neighboring villages. Stories offer a richness of place, and storytelling plays a significant role in creating experiences, forging identity, and crafting memories of villages and people. Living in the shadows of a carbon landscape where names of towns and villages mark oil and coal activities, oil fields, and signposts, storytellers along the foothills express place-making through distinct accounts of social relations and history (Raffles 2002: 28).

Today, hundreds of police checkpoints in the foothills operate as a border between Assam and Nagaland and mark the presence of the territorial states. It is impossible to ignore the boundaries that delineate every piece of land and village here. Houses, fields, villages, and tea plantations that stretch across the landscape are all fenced. So are the oil rigs and hydrocarbon exploratory units in the lower elevations, and the coal mines in the Naga villages higher up. Bamboo poles demarcate individual homesteads inside the villages, lines of mud walls divide the rows of rice fields, hillocks serve as borders for *jhum* fields, gravel roads divide some villages, and columns of trees and stones mark the boundaries of many more villages. Geologically called the Naga Schuppen belt, the slopes, rivers, and plains on which these villages and towns have arisen are recognized as the site of rich hydrocarbon deposits. This hydrocarbon "belt" falls within the territorial and administrative jurisdiction of Nagaland, Assam, and Arunachal Pradesh, causing routine land conflicts. As a result, some people appear unsure during conversations about where the official border lies. "Some say that it is the tamarind tree outside the village, others say it is the last house on the hillock.

I do not know," commented Bahadur, a Nepali resident who lives along the Assam-Nagaland foothill border.

OLD TSSORI AND NEW TSSORI VILLAGES

We were sitting outside his kitchen one winter morning as Patton narrated the history of New Tssori and Old Tssori. Both villages had strong ties with the *haats* (weekly markets). Most Naga traders at the Rajabari and Nagabat *haats* came from these villages. According to Patton, New Tssori village was an offshoot of Old Tssori. Patton's great-grandfather was one of the founders of Old Tssori. The history of Naga villages in the foothills is often connected to old villages in the upper elevations. It was common for kin groups from particular villages to migrate and establish new villages for reasons such as internal disputes between clans, exploration of new lands for cultivation, or diseases and epidemics in the old settlements.

Many villagers living here said that they migrated either from the uplands of Nagaland or from the Brahmaputra Valley in Assam. In some cases, residents had moved from an older village in the hills but maintained their kin connections and ties with their clans there. This was the case with Patton too, who shared the history of two Lotha (Naga) villages, Old Tssori and New Tssori, both of which had been recognized as important sites for hydrocarbon exploration by the government of Nagaland. The community lands of both villages also featured in the ongoing contestations and debates about the Naga people's aspirations for a carbon future in the hills of Nagaland. "Old Tssori village was established in 1865 and New Tssori village in 1955," explained Patton.

According to Patton, the settlers of Old Tssori came from Akok village in the hills. Old Tssori had been established in the nineteenth century "under the British rule" because "[our] forefathers wanted to live close to the plains. The main reason was the attraction of the wet rice cultivation and access to the markets in the plains," Patton said. But these pulls were not new. Inhabitants living in the surrounding Naga Hills had maintained strategic relations with the Ahom kingdom since precolonial days. Returning to the story of setting up Old Tssori, Patton said that for the first ten years (1865–75) the village functioned without the knowledge of the British administrators, because "the residents did not know that the village was supposed to be registered with the administration." Therefore, the official documents of the village show its founding year as 1875, but according to Patton the village dates to 1865. He explained that New Tssori was founded "the year I was born" and gave an account of Old Tssori and New Tssori by invoking kinship ties and internal migration.

New Tssori was started because of a dispute over positions of leadership in Old Tssori. "At that time, there was a post on the bench court in every village. The British government had established this post for us. The bench court chairman had to work in consultation with the administration to carry out activities in the village." After India's independence, the system of governance in the Naga villages remained unchanged. Two members from his kin group competed for the post, and due to the dispute Patton's family moved out. Patton's parents migrated to New Tssori in 1956 because they wanted to improve their lives by settling close to the foothills, where they could have access to both the hills and plains. "The land in the foothills was so fertile that they used to load nine or ten bullock carts of vegetables during the harvest season," Patton said. But after the establishment of the hill state of Nagaland in 1963, a sizable portion of the rice cultivation areas of Tssori village fell under the territorial jurisdiction of Assam. This issue of losing village lands to Assam often came up in our conversations.

Patton insisted that the conflict which had divided the old and new village was an "internal matter" and not a "public issue." To reestablish the official story of Old Tssori and New Tssori, he picked up a stack of laminated documents he had neatly arranged on the table earlier that morning. As though he was adjudicating an important case, his language became administrative and focused on a colonial lexicon about boundary demarcation and land relations in the foothills. The process was used to distinguish what he defined as the official version from the unofficial or internal matters of his village.

With an air of urgency, he ordered me to "write down," as he put on his reading glasses, lit a *beedi* (hand-rolled cigarette), and began to read from the laminated document: "1883 is the Doyang Notification. Total area is 60,865 *pura*. July 1885 Notification—Doyang—this is the name of the reserve." He read out the official allotment of land, flipping the laminated pages to connect the story. "After this first notification, the first pillar of this area was put up in the Woka Goronga tea garden," he announced and flipped some more pages. He regularly paused to interpret the documents. "Name of valley—Desoi valley touches the Mokokchung and Longleng districts in Nagaland. The first notice was 1st November, 1883." I was unable to connect the sentences he selectively read out from the documents or follow the text and the explanations he provided, but not wanting to interrupt him, I continued to record the details. A few minutes later, Patton paused, drew hard on his *beedi*, and passed the laminated document to me, saying, "Read the

Patton with his laminated documents

agreements and write down the details." Eager to see what kinds of documents he was citing, I took the papers. They were British colonial notifications that had appeared in standard colonial gazettes and ethnographies I often come across in bookshops and libraries.

Patton commented that the official inscriptions were flawed, with errors in the names in the official British documents about his village, including the spelling of his great-grandfather's name on a village notification letter dated around 1875. According to Patton, his great-grandfather was named Nymtsamo, which meant "oppressed by others." But it was misspelled in the colonial document as Yimtsamo, a name with no significant meaning. Patton said such errors were common because "there was no writing in the village." Although his great-grandfather was the village headman, he did not know how to read and write, and therefore was not equipped to correct the error. For Patton, it was important to correct the names of his family members and the villages as a step toward rectifying the meaningless names in the official historical documents.

Personal names have social value and play a significant role in constructing identity. Names can both establish and erase a person's social status and provide a key to other important information such as gender, religion, and kinship (Bruck and Bodernhorn 2006). Similarly, naming people and places stabilizes identities and aspirations (Hansen 2006: 201–22). In Patton's case too, correcting names of people and places in the colonial documents was crucial, because unless the errors were rectified, Patton felt there was risk of "losing all the disputed lands to the government of Assam."

His aim in correcting the name of his great-grandfather was to authenticate the history of his villages and claim a legitimacy of belonging to the land, which was increasingly gaining attention as a hydrocarbon reservoir and the site of future oil fields in Nagaland. The increasing number of visits by government officials, ONGC officials, traders, and politicians to inspect the village lands only seemed to instill a sense of urgency to provide the correct histories of the villages. For Patton, the colonial documents not only served as a reminder of the official history of his village and the boundary of his ancestral lands but also documented the injustices that British administrators had carried out in his community. Shuffling through the stack of files, he handed me a colonial document that showed the important role to which his great-grandfather had been appointed—as a tax collector for the British administration. As I read, he selected another set of documents as additional proof of the history of his village and the land, and passed it to me.

Patton preserved the documents neatly in colorful files, taking care of the photocopied documents, which contained the history of his village. Today, huge tracts of the cultivable lands that the Tssori villages (Old and New) once held in the foothills fall under the jurisdiction of Sibsagar district in Assam, a district marked with oil wells and geological explorations. It is not hydrocarbon structures but a rural market known as the Rajabari *haat* that the people of the Tssori villages recognize as the marker of their history. Patton's family believes that some of the Tssori people's lands became part of the Rajabari tea plantation. Initially the Tssori villagers visited the Rajabari weekly market, a small *bagan bazaar* (tea plantation market) meant for the plantation workers, which eventually became a weekly *haat* open to people from surrounding villages. Patton explained, "Until today the tax collectors do not collect tax from the Nagas in the Rajabari *haat*. The plains traders are all taxed. This is because Nagas allowed the market to be established on their own land. That's why they are not taxed." This was not exactly true. During my fieldwork, when I asked why Naga traders were excluded from paying taxes at the weekly Rajabari market, the Ahom tax collectors told me that the Nagas were extremely poor and the produce they brought from the hills did not fetch high profits. The tax collectors added that they made an exception for the Naga traders at the market because of social relationships between the hill people and the valley people.

When I met Patton, he said he was a *lichon* (cultivator) and suggested that we sit outdoors. He arranged two cane stools and a wooden table where he organized laminated documents and folders for our meeting. What had seemed a simple research interview for me was an important occasion for Patton. It was an opportunity to narrate the story of his life and the history of his village, and to display the colonial documents he had collected over the decades. He was creating an official atmosphere. After I helped him carry the files indoors later that day, my attention was drawn to a 2010 *jhum* cultivation calendar hanging on the bamboo wall. The months, weeks, and days were scribbled with notes, numbers, and diagrams. Taking a closer look, I noticed that the calendar also functioned as his diary. The activities he had to attend were listed there: vaccinating the pigs, attending to his *jhum* fields, meetings with local church members, and sowing and weeding. The calendar on the bamboo wall presented a snapshot of Patton's life in the foothills. Patton's self-appointed role as record keeper and storyteller explained the numerous meetings he dutifully attended. Many villagers, whether Naga, Nepali, Adivasi, or Assamese, told me about losing lands to their neighbors, or were involved in negotiating disputes over lands that were categorized as potential oil reservoir rocks and thrusts (geological formations that trapped oil). The story of Old Tssori and New Tssori was not exceptional in that regard. In other cases, people invoked the presence of the state to create a legitimate history of the village.

LONGTSSORI VILLAGE

It is not wise to travel to Longtssori village during the monsoon. After the tires of the jeep sank in the red puddles, we began to walk along the wet red dirt road toward the village. The first house I entered in the village was the pastor's residence. A large crowd of men, women, and children had gathered there for a free medical checkup as a team of doctors and medical assistants sat at a table distributing antimalaria tablets and vitamins. "It is a mission," one of the volunteers told me, explaining that none of them represented the government. Since there are no clinics or schools in Longtssori, churches and cultural organizations routinely ask individuals from various departments to volunteer in the village.

Days after the crowd had dispersed and it had stopped raining, the village appeared to be washed in a haze of green. The forests, tea plantations, paddy fields, vegetable farms, and kitchen garden gave a pleasant feeling to the village. Up to this point, I had not seen the new coal mines beyond the village mountains. What was it I wanted to know about this village? Dates? Stories of origin? Evidence of falling in a disputed zone? I was unsure where to start, yet the conversation with the pastor seemed to take its own course.

"To start a village, we go around and inquire about the place, and feel the water and air of the place. But it was not the case with this village," said Jami, the pastor of Longtssori village. The residents had migrated from a village in the higher elevations called Liyo Longitang, and the name Longtssori meant "slippery pebbles" in the Lotha language. The village had been officially recognized by the state of Nagaland in 2007. Residents explained that the location's main appeal was its accessibility to the markets and wet rice fields nearby. Although people from the village had been cultivating the land here since 1937, they lived in the upper elevations. Jami said, "We were in the mountain village. We would come down to cultivate and go back. But we decided that it no longer made sense to travel up and down between the foothill rice fields and our village in the mountains." When old cultivators, unable to walk long distances to the hilltop villages, started staying in their wet rice fields in the foothills, discussion began about establishing a new village.

There are regulations and procedures to be followed before officially recognizing a new village. For instance, government officials visit the village site to verify certain official indices. To apply for government recognition, the village has to contain at least thirty households. A government circular is then sent out to determine if neighboring villages have objections to the new village. If there are no objections to the boundary and other matters, an affidavit is drawn up and official signatures are collected from representatives of the neighboring villages. Finally, the district commissioner of the state (in this case, Nagaland), with the approval of the governor of the state, officially recognizes the new village in the state gazette.[1] Once a village is recognized, state benefits such as schools, clinics, government offices, police stations, banks, post offices, village councils, and customary courts are sanctioned, but these services remained on paper for Longtssori village.

"We were lucky at the time. My younger brother was a government bureaucrat, so he pushed our case. Others have been seeking government recognition but have not been successful, but we were lucky because of my brother," Jami said, explaining why paperwork for the village moved quickly. He reflected on his experiences with the paperwork and his brother's role in pushing through the approval: "People say government does it, but it is the people who constitute the government. The people might say governments carry out work, but it is people who do it."

Following the village's official recognition, residents decided to establish a church. On October 12, 2008, the new church was inaugurated and a pastor was appointed. There were two village headmen and a village council too. Residents of the village said they were adapting to the landscape and as a result had become lazy. When I asked them what they meant, Jami said, "We do not want to climb the hills anymore, we cannot climb the hills, we get tired easily." The villagers' main concern was the lack of a clearly marked border between Nagaland and Assam. Conflicts over the territorial demarcation in the foothills haunted them. In 2006, some groups from Guwahati in Assam entered the village's rice fields and started to erect a stone wall. Residents of Longtssori stopped them and demanded to see the government order and permits. When the Assamese group failed to produce any official document, Jami and his friends explained official procedures to the fence builders: "If you want to build anything here, it should have the consent of the Nagaland government or at least a signature from the border magistrate. If there is a plan to install border pillars [concrete boundary markers], there have to be officials from both sides—Assam and Nagaland state officials representing the governments have to be present. But in this case, the people installing the pillars were from Assam, but there were no representatives from the hill government, so how could we allow them to put up the pillars? They were lying. They were humiliated and scared."

Documentation is important. But residents of Longtssori organize activities such as mobile clinics to show that the village is part of the state administration in Nagaland. Given that everyday economic activities are frequently marred by land conflicts between the Assamese and Naga villages, the residents of Longtssori organized official functions and health camps to showcase the presence of the state in their village. As a result, residents frequently boasted to their Assamese neighbors that state officials from various departments had visited the village. Thus there is a strategic invocation of the state, although basic state amenities such as roads, schools, clinics, or government offices are absent. Residents take care of themselves by finding work in coal mines, rubber plantations, tea plantations, logging, or by growing cash crops such as yams, ginger, areca nuts, and betel leaves.

The revenue from the coal trade and tea plantations on village land belongs to individual traders and households. Eventually, I met young traders from the village who were excavating a new site for a coal mine in the upper elevations of the village. The workers in the new coal mine came from neighboring Adivasi and Ahom villages in Assam. I noticed that the soil was gray, and a foul smell came from the cave. "There is coal here," announced the excited Naga traders, yet work was going slowly. Only shovels and spades to dig the new coal mine and around a dozen jute sacks for carrying out debris lay next to a heap of mud. "We will get the coal," they said. Inside the mine, the exposed soil was moist and warm, and the fumes burned our eyes and nostrils. The colors of the walls in the cave ranged from brown to dark gray. A perennial spring nearby, disrupted by the digging operations, overflowed the pathway to this new mine and created shallow pools of red water around the site. In the mine, conversations about routine conflicts in the foothills came up again. One of the Adivasi workers said, "Diktar ase" (It is difficult), referring to the challenges of earning a livelihood under the prevailing tension.

The "border problem"—the boundary dispute between Assam and Nagaland—is a common theme. In Longtssori and the neighboring villages, what is defined as the border problem often turns out to be everyday experiences of residents who constantly cross one another's social and physical spaces, such as village lands and community forests. Some residents were annoyed and called these crossings encroachment instead of trespassing. Any person from a different ethnic group entering the land or the wet rice fields of another ethnic group is viewed as a potential encroacher. Everyone is suspicious. "The people from the nearby villages in the plains [in Assam] freely come to the foothill areas and take away resources such as timber, vegetables, and other things—they do what they feel like doing. They bring their cattle for grazing in the foothills," explained Aben, a cultivator from Longtssori. Stories about losing land to Assamese people were common. Likewise, villagers who fell within Assam's jurisdiction said that Naga villagers in the foothills were involved in dubious land transactions and were constantly encroaching on lands that belonged to the people of Assam.

Around Longtssori, descriptions of land encroachment are vibrant and innovative. I commonly heard references to an iron chain, emphasizing the presence of a rigid border. However, there was an ongoing battle over the boundary. It was as though two parallel lines—the Assamese and the Nagas—stood on the land, each holding an iron chain (to signify the idea of the border as a straight line) and pushing each other to stay in their respective places and also gaining more land any time they could exert their power: the Nagas toward the hills, and the Assamese people toward the Brahmaputra Valley of Assam. In a similar fashion, cultivable farms were also described as suffering from these conflicts. "Imagine a foothill field with a zigzag boundary," a Longtssori resident told me as he explained how farmers from Assam had encroached on the Naga village lands. In his imagination, the foothills appeared like a straight line that disrupted the hill/valley boundary. However, the same argument appeared in the Assamese villages as they stressed that Naga villages in the lower elevations of the foothills were illegal settlements and that tribal people belonged to the hills. People from Assam and Nagaland explained that whenever they were pushed off their land, their respective governments intervened to put them back. These accounts of establishing new villages and starting new tea plantations represented the states in the foothills as entities that distributed "empty" lands to their respective citizens in Assam and Nagaland.

"So you can imagine. How can one individual fight against nine or ten farmers from the plains [Assam] who are pushing together toward the foothills and beyond?" Jami said as he reflected on the land conflicts with Assamese villages. Feelings of being overwhelmed by the people from Assam permeated relationships and imaginations in Longtssori. Residents insisted that people from Assam even cast magic spells on activities such as trading and cultivation in the Naga villages. To ward off these spells, my host and his wife made paper crosses and wrote down passages from the Bible, pasting them on doors and windows. I noticed paper crosses with inscriptions such as "Oki shilo kuri ji jisu" (Christ is the head of this house), "Kipang shi cho Jisu na nhyakala. Kipvui ji jo Jisu" (Jesus stands guard at this door, and the owner of this house is Jesus), "Satan nina opvui ni potsow khuma" (Away from me, Satan! For it is written worship the Lord your god, and serve him only). My hosts described how these religious signs kept away impure charms and black magic practices that were rampant in the area. Emphasizing the anxieties about living in the foothills, my hostess exclaimed, "No one from here will get to heaven! It is impossible to remain calm and at peace in such places." When I asked her why that was the case, she explained:

> First, it is impossible not to be angry at the plains people [non-Naga groups from Assam]. Look at the thieves who come and steal from our house—they take away our pots and pans. It is impossible to keep anything outside. Second, look at the way they come to the hills and take away our vegetables, chop down our bamboo groves, and carry away our trees and forest resources. They make us sin. We are always angry at the valley people's attitude and their activities. Thirdly, look at the kinds of black magic—the spells they cast on the hill people [the Naga people]. My father-in-law died of a black magic spell.

The pastor described the black magic spell that was cast on his father: "There were four black magic spells cast on my father. I managed to cut away the first three spells, but the fourth one was very powerful. It killed my father." Explaining the fourth magic spell, he recounted that the spell was cast on his father's bed. As a result, his ailing father was constantly frightened of his own bed and refused to lie down there. Instead, he jumped up, waving his hands in the air like a child for hours at a stretch, and passed away shortly after the spell had consumed him. They also cautioned me about the dangers of coming under an evil spell and specifically warned me not to touch strangers. "When you meet people, just fold your hands and say, 'Namaste,'" they advised soon after I arrived. According to them, the *namaste* greeting, a standard way to greet people in the plains of Assam, did not require touching and was therefore a safe practice.

Given the anxieties about living here, strangers or *oyem* (others) are feared figures. The sociological concept of the stranger is created by the tension between freedom and fixation to boundaries that lingers in every relationship between the stranger and the residents of a place (Simmel 1971: 145–48). Yet, in Longtssori, the stranger is not defined by mobility but has a fixed location as someone from Assam. Unlike the figure of the stranger that Simmel describes as objective, nonpartisan, and free from social entanglements and prejudice due to his mobility, strangers from Assam have vile intentions. They enter the village for various trading or business purposes but surreptitiously survey the land to extract resources and encroach on it. Therefore, Longtssori residents always try to establish a relationship with strangers who arrive in the village. They talk to them and locate them within a specific relational network, such as a kin group, an ethnic group, a caste hierarchy, a village, or a district.

For Naga villagers, the strangers are the Nepali, the Bengali-speaking Muslims, and the Adivasi people from Assam.[2] These are perceived as poor people who arrive seeking land, employment, and resources. Stories about the Adivasi people as a community "hungry for land" circulated in the Assamese and Naga villages. Tribal customary law permits only Nagas to possess land in Naga villages. However, given the nature of the land conflicts, overlapping claims to land often create tension and lead to violence between villages. Therefore, conflicts between Naga villages and non-Naga settlements quickly become border conflicts. This is further complicated by assertions of ethnic purity. Assamese and Naga villagers alike deny that ethnic groups such as the Nepali, Bengali-speaking Muslims, and Adivasi belong in the foothills.

GOREJAN VILLAGE

I met many Adivasi people who worked as sharecroppers, tea plantation workers, and coal miners. Most of them were daily wage laborers making ends meet, yet many asserted that they lived in the *gaon* (village) and not on the tea plantations. These departures from the tea plantations to the villages were anything but ordinary in Assam, resulting from labor protests on the tea plantations of Assam between 1900 and 1930. The British administration called these departures desertion, and Adivasi workers caught leaving the plantation were arrested and flogged (Behal 1985).[3] According to historian Rana Behal, it was the most common way to escape the harsh plantation life and was a form of protest against the various kinds of extralegal authority that were imposed on workers.[4]

There was heavy traffic on the paved concrete road that cuts across Gorejan when I first visited the village in 2009. Heavy trucks and tankers from the ONGC gas gathering station at Borholla pass this way. Installed in 1981, this is one of the oldest ONGC units in Northeast India and continues to be marked as an area with a "high potential" for oil and natural gas resources. As a result, in addition to the ongoing exploration activities, new proposals to drill wells and conduct hydrocarbon explorations here remain a priority for the ONGC (Asian Conssulting Engineers 2016: 1). Just outside the entry gate, the road splits in two. One path leads to the Rajabari weekly market where residents from Old and New Tssori regularly come to trade. The second path goes to Champang ONGC village, where abandoned oil wells and pipes leaking crude oil stand in the middle of village land. The village is located close to the oil and gas wells, and the establishment of the ONGC station has become part of village history.

Gorejan is partly hidden by paddy fields and tea plantations. Only the whitewashed Catholic church stands as a significant landmark. Augustine's father, Samar, mentioned that oil exploration around his village started in 1970s, thus dating the infrastructure around the village, which includes culverts, paved roads, and electricity. He recollects running out of his house as a young man to watch trucks and machines going to set up the ONGC hydrocarbon base. His parents visited the oil site to watch the drilling activities. Soon after the oil exploration operations started, workers and contractors arrived to construct roads, culverts, and bridges around his village. Then the traffic increased as the large trucks brought in equipment for the oil rigs.

Gorejan village was established in 1933 with ten Adivasi households. "The Assamese people followed us here; they were afraid to come here," Samar said. "But the Assamese households increased in the village, and today they are a majority." When I asked him where he paid his land tax, Samar responded, "Once we got the *patta* [land deed], we paid the tax to the British government. Now, it is paid to the government of Assam. But the Naga villagers pay their tax to the government of Nagaland." When Samar's parents came to Gorejan, they brought a Catholic church to the village. In 1977, they constructed the concrete church that stands as an important landmark of the village today. Approximately eighty households belonged to the Gorejan church. Samar's memories of the new village, the church, his paddy fields, and the forests in the area are all tied together.[5] "I worked on the soil, [but] I was also working in the church," he said, adding that he served as a catechist in the Gorejan church for twenty-eight years.

For Samar, the church was an important part of his life there. All ten of his children attended the Catholic schools around Gorejan; the eldest son worked for the church, and his youngest daughter was a nun. Jacob, Samar's eldest son, said his teacher in elementary school was an Englishman who had offered to take him to England, but he thought, "English boys and girls will laugh at me since I am an Adivasi boy." That was in 1968. Although he never visited England, Jacob went to Jharkhand, which he referred to as "our homeland." Like ethnic groups in the foothill villages who trace their relations to Nagaland, Assam, Nepal, and Bengal, Samar's family traces its kin connections to Jharkhand.

These connections are best captured through the various cultural and political alliances people maintain here. For instance, Samar's son Jacob explained why he kept memberships in both the Assam Tea Tribes Students' Association (ATTSA) and the All Adivasi Students' Association of Assam (AASAA). ATTSA membership accommodated both the tea tribes (plantation residents) and the ex-tea tribes (those who had moved off the plantations), but AASAA was an organization only of ex-tea tribes. He said, "We have to be in ATTSA and AASAA at the same time. We have to be in AASAA since we were born in Assam, but we are tribal people in Assam so we also have to be in ATTSA. Because they are the same people but different organizations, the government does not listen to us." Jacob's alliances also extended to his ancestral home in Jharkhand. Most Adivasi Christians in the area were members of the Roman Catholic denomination and shared a common ancestry with the parish priests and nuns, who came from Jharkhand. AASAA's own organizational history was tied to the parishes of upper Assam as much as it was to those who aspired to leave the tea plantations to seek a better life elsewhere.

I learned that Jacob had visited his ancestral homeland in 2009. He said, "I went to Jharkhand to visit my father's family. It was in the Dorma area in Khunti district." According to Jacob, the place was named after their grandfather, who was called Bari. He stayed there for one and a half months with his uncle Petrus. On the last day of his visit, Petrus cut a ripe jackfruit for him. Jacob continued, "There is a jackfruit tree in his house. Uncle said that my great-grandfather planted that tree before he left for Assam. I enjoyed seven pieces from the jackfruit." Eating the fruit and carrying the seeds back to Assam held immense significance for him. The jackfruit tree in Jharkhand, the sweet taste of the fruit, and the story about the journey of the seeds, captures the story of origins, land, and the afterlives of indentured labor in the foothills.

Jacob's story of the jackfruit tree and its seeds is not an account of displacement. Instead, it highlights how people construct and lay claim to new places by naming them, planting seeds in the ground, and tilling the land. It illustrates the power of storytelling and the experience of politics and relations. Jacob recounted the history of his family with pride: "At least I ate something my great-grandfather planted. The jackfruit was so tasty. I cannot even describe the taste. It is the sweetest jackfruit in the entire Munda area. The taste of the jackfruit was so legendary in the area that people walked from faraway places—as far as twenty to twenty-five kilometers away—to eat the fruit."

He repeated the story of the jackfruit and added, "My great-grandmother also planted a tamarind tree. It was God's plan that I visit Jharkhand." Jacob's father, Samar, sat with us listening to the story. When I had met with Samar earlier that day for an interview, I suggested that he could talk to me in any language he preferred, Assamese, Hindi, Sadri, English, or Nagamese, but he waved his hand to dismiss my offer: "Nai, moi karona Nagamese thik ase" (No, for me Nagamese is all right). I realized that the everyday choices people made, such as the languages they embraced, the friendships they forged, and the villages they called home, gave them a deep sense of belonging and meaning. "I was born in Nahorali Bagan [a tea plantation] near Lakwa train station in 1929," Samar said. He was born on a tea plantation, but his family decided to move away when he was four years old.

Historian Jayeeta Sharma (2009) describes how indentured laborers who had been brought to the tea plantations of Assam in the nineteenth century felt they had reached the end of the world. Many workers died in the hostile environment of the Assam tea plantations. The feeling of loss and alienation was deep as they were forced to work under extremely harsh conditions. The making of the plantation coolie in colonial Assam, as Sharma notes, involved a violent process of controlling and disciplining the labor population to optimize the production of tea in the Brahmaputra Valley.[6] From 1870 to 1900, the number of indentured workers recruited from eastern and central India was estimated between 700,000 and 750,000.

Describing the experiences of indentured workers in the Assam Valley, historian Rana Behal highlights how the adult labor force was subjected to "strict control through penal laws, floggings, illegal confinements, and the *chowkidari* system. Disease, malnutrition and a high rate of mortality were the harsh realities of plantation life for the labourers. . . . This attitude [the employer-employee relationship in the plantation system] was akin to that of the white masters towards their black slave labour in the antebellum era in southern USA" (1985: 19).

Today, descendants of the indentured workers in Assam are known in Northeast India as Adivasi. Although Adivasi literally means "original inhabitants of the land," in Assam it has increasingly become an ethnic category. Adivasi people mobilize and assert their political identity based on shared cultural and social experiences, including language and history.[7] For residents like Samar, the foothills are a communal place where he grew up and raised a family.

As Samar narrated the story of the village, he suddenly smiled and said that even Japanese and American soldiers had come there. With an intimate familiarity, as though the incident had taken place recently, he stretched his hands to point and said, "Their vehicles got stuck in the mud over there. They camped for a week and left. Everything here was a forest." He was referring to World War II battles fought on the frontiers of Northeast India. For Samar, the foothills were an important place where global events like the World War II battles once took place.

When I asked why his parents had decided to settle in the foothills, he said, "Because we caught land here." He made a gesture of catching an object and explained that land is slippery in the foothills. Samar's account of securing land puts in question notions of land as being always demarcated from the beginning.[8] It also suggests how social relations that are central in establishing boundaries can be obliterated, made invisible, or swallowed up by terms like *fuzzy* and *interstitial*. These terms seem to imply that everyday human experiences are fixed from the beginning within a rigid national map, and with each crossing, lines and boundaries become blurry. Samar's account of land reiterates Hugh Raffles's reflections on place-making. Raffles notes, "The conflictual and ongoing work of place-making is here expressed through the idiom of shared pasts. Stories of this type call on nature to reinforce belonging, and they anchor place, yet dispute its meanings" (2002: 55). On the one hand, land is perceived as an independent entity that must be secured and tamed by investing one's labor and time. On the other hand, it is also the foundation where the interests of the community and notions of sovereignty and belonging can be planted, like the story of the jackfruit seeds from Jharkhand.

People's imagination and accounts of "catching land," whereby land is given agency as being alive and slippery, allows us to reexamine the epistemological grids regarding notions of geography and places. Samar's sense of geography was marked by his memory. It stretched from the eastern Indian village from which his ancestors had been brought as indentured workers, to the tea plantation in Assam where he was born, and finally to the foothill border of Assam and Nagaland that he called home. His memory of the place was populated with friends and neighbors including the Naga and Assamese people he had met when he arrived here. In addition, a political and moral dimension of time informed the Adivasi sense of belonging and place-making. The past was neither glorified nor rejected. During long conversations with Adivasi residents around Gorejan village, accounts of the past were invented and constructed, and at other times altered and arranged in a chronology, much like other communities who lived along the foothills. Yet their accounts of home were distinct because they lacked a typical ethnic history that was well crafted and presented as a standard account. Instead, Adivasi families frequently invoked friends and neighbors in their stories, unlike their neighbors, who often claimed that they were the "first" to arrive there.

According to Samar, his friends from the neighboring villages came around the time his family arrived in 1929. Landmarks, forests, and memories of plantations and people frequently came up during our conversation, reflecting how, as William Cronon aptly noted, "the cultural values of people as storytelling creatures struggling to find the meaning of their place in the world . . . carry us back and forth across the boundary between people and nature to reveal just how culturally constructed this boundary is" (1993: 32–33). Drawing from his memory, Samar gave a list of villages around Gorejan, mentioning people and landmarks to highlight that it was already an inhabited place. Furthermore, when he talked about transformations around him, such as the disappearance of the forests or the establishment of tea plantations, he described the changes through his daily activities. For instance, he told me, "Nowadays, we have to travel really far into the hills to get bamboo. Before, if we went as far as Kithan's house, the bamboo was very strong and thick. They were so long that both ends of a bamboo would touch the ground when you carried it over the shoulder."

Samar's recollection of the bamboo groves captures the essence of belonging and his relationship with the land. He insisted that I visit people in the neighboring villages to reconfirm events that had occurred around the place. Like oral footnotes, he repeatedly paused and commented, "Ah! For this story, you must go to Kithan and ask him how we carried out the activity." Or he would say, "Oh! For that story, you must go to Gogoi." It became increasingly clear that the foundation of an inclusive or exclusive history was recognizing or obliterating one's neighbors, friends, and alliances in the foothills.

CONCLUSION

Contrary to the perceptions of state officials in Assam and Nagaland, or Indian security forces who view the foothill border through the prism of law enforcement, for residents it is a place where social histories and communities change every three or four miles. This is evident in accounts from the inter-ethnic settlements of Naga, Ahom, Nepali, Adivasi, and many Assamese villages. However, narratives about the foothills as an ethnically diverse or uniform and rigid landscape also depend on the speaker's political and social orientation.

The process of storytelling presents the multifaceted relationship that communities forge with the land and their neighbors. Access to land and resources is central to establishing a political identity here, and these in turn shape what constitutes the history and memory of the people who live here. It is land that enables communities to make different kinds of claims of legitimacy and belonging. Everyone along the foothill border between Assam and Nagaland came from elsewhere. Every community has a story of arrival. As they share experiences about belonging, they continue to create a history that is produced every day in the absence of a recorder, an archive, or photographic documentation.

CHAPTER TWO

Difficult Loves

A POPULAR way of telling stories in Nagamese (the lingua franca in the foothills of Assam and Nagaland) is by invoking *morom*, a broad term for love and affection. *Morom* plays an important role in determining social order. As land, coal mines, plantations, fields, and business networks are handed down through a patrilineal order, it is the tribal councils and associations that decide the rights, obligations, and honor of the larger community, even to the extent of determining which practices are deemed acts of *morom*. Acts of violence or affection in intimate relationships are validated and socially constructed as social models to be upheld or rejected by these groups. In this way, the process of categorizing marriages and intimate relationships becomes a kind of theater whereby the whole village or town is involved in passing judgment or commenting.

Morom is distinct from the English word *love* in that it encompasses all kinds of attachments and affections. It is not isomorphic with *love* but resonates with everyday articulations and declarations of affection. *Morom* captures multiple emotions and relationships, and is not limited to romantic love or affection. It refers to relations of patronage between servant and master; bonds between friends; attachment and caring between parents and children; relationships between the state and the public; and lust, attraction, and adoration between lovers. Mercy, gratitude, sympathy, grace, compassion, and charity are also encompassed as acts of *morom*.

In the foothills, affection and love are framed to iterate a cultural and political understanding of territoriality, violence, and power. In particular, kinship and gender relations in the coal mining villages were often described as acts of *morom*, but this rhetoric was strongly grounded in a politics of ethnic purity that was often mapped on women's bodies. Given that land and resources are central to establishing ownership, power, and authority for communities and individuals alike, and resource conflicts are routine in the foothills, the classification of *morom* are most visible in the gendered politics where women's bodies are presented as sites of purity, danger, primitiveness, and promiscuity.

MY HEART WILL BE EMPTY WITHOUT YOU

I started thinking about love during my fieldwork in 2010, one night after a long, tiring day of interviews in the foothills. The household where I was staying had three members: a father, a mother, and their son. All the other children had left. Some had married and settled in neighboring villages, while others had gone to the city to pursue studies and seek employment. I was given a room that had belonged to one of the children. It was the end of November and the nights were cold, so I added my shawl to the blanket that was given to me to keep warm. As I looked around the room, my gaze lingered on a dressing table lined with old albums, empty cream and powder containers, and small plastic decorations. The dogs were howling on the street as I lay in bed tired but unable to fall asleep.

A heart-shaped plastic decoration hanging on the wall caught my attention. Along with plastic flowers and other items, such as empty containers and photo frames, this slightly faded plastic stand with the inscription "My heart will be empty without you" stared at me. Had this been given to a child at school? Was it a gift from a first love? What would it be like to fall in love in a place like that? These questions stemmed partly from frustration, exhaustion, and the bitter cold but more from my experience of the town.

"Where is everyone?" I wondered one afternoon as we drove into town. There were no signs of life, except for the street dogs barking and chasing each other. All the shops were shut, and everyone was indoors. "These are bad days," my hosts whispered as they instructed the driver to keep his eyes and ears open and to guard the jeep in which we traveled. Later that night, the host's son recommended that we unplug the jeep battery and bring it indoors, as there were thieves and drug addicts who picked up vehicles, water pumps, and anything they could lay their hands on.

I learned that my arrival in town had coincided with that of two Naga insurgent groups. Since the 1950s, Naga insurgents in the hills have waged a protracted war with the Indian state, demanding the right to self-determination (Kikon 2009; Lotha 2007).[1] Between 1975 and 1989, the armed movement split into three factions, which led to violent conflict over control of the hills. In the protracted armed conflict across Northeast India, the language of devotion and sacrifice has been invoked to encourage support for a nationalist sovereign homeland,[2] which is most visible in the claim to hydrocarbons as resources belonging to the people. In Nagaland, this is marked by increased pressure to come up with a plan for oil and natural gas explorations in the state and to take over the coal mines in Naga villages. In Assam, the debate over unpaid oil royalties is a key point in defining the extractive regime in the Brahmaputra Valley as an enclave economy in post-colonial India.

My host described how residents were caught in a battle between two powerful armed groups—the National Socialist Council of Nagaland (Isak-Muivah), commonly referred as NSCN(IM), and the National Socialist Council of Nagaland (Khaplang), known as the NSCN(K). The conflict involved the right to collect the "national tax," in which insurgents levied taxes on the public and collected them in the name of Naga nationalism. Ultimately the two groups came to an understanding. Every three months, there was a transfer of power between the NSCN(IM) and the NSCN(K) to collect the "national tax" and control the town and its surrounding villages. During such power transfers, there was a danger of misrepresentation as there were several factions of the Naga armed groups, each deeply suspicious of the other. Therefore, during such changes of the guard, the new group taking over power stopped vehicles to interrogate new faces and monitored movement on the streets. So it was best to remain indoors.

The Naga insurgents who had arrived in town during my stay professed nationalist love for the Naga homelands and the Naga nation.[3] In the foot-hills, there were competing sovereign groups that ranged from state actors such as police and border magistrates all the way to the cultural associations and of course, nonstate actors like the NSCN, all of whom asserted their power and authority over the villages and towns. Insurgents also regulated and controlled the coal mines in the Naga villages. Territorial boundaries were drawn on ethnic bodies, and people reiterated the essence of Naga-ness as a body that belongs to the hills or Assamese-ness as a valley identity.

This way of categorizing ethnic bodies determined the nature of relations and alliances in resource extraction operations. For example, there were few land conflicts around the Naga coal mining sites, where customary law guaranteed that the Naga people owned the land and natural resources, unlike in Assam. Assamese traders and Naga landowners worked as partners and schemed over ways to increase the coal mining operations and seek other business ventures. Therefore, there were cases where Assamese traders helped Naga villages ensure that lands marked for coal mines would come under the jurisdiction of Nagaland so that the trading parties could extract individual profits and keep the coal mines outside the purview of state law in Assam. But this was not the case with tea plantations or rice fields, which were privately owned by ethnic groups from Assam and Nagaland. Thus, numerous conflicts and deaths occurred in the disputed lands over overlapping claims to further establish tea plantations, farms, or settlements, but seldom over coal mines.

Every time Naga residents move to a new site to open a coal mine, or Assamese people arrive in the foothills to start a plantation, or the ONGC finds oil in the foothills of Assam, geographical boundaries and human settlements are reshaped. Issues regarding ownership of land and resources are important to many ethnic groups who live here. Historically, ethnic groups on India's northeast frontier have mobilized for political recognition based on exclusive cultural identities rooted in specific geographical areas.[4] Today, demands ranging from special constitutional status and autonomy all the way to new ethnic statehood continue to be founded on the logic of exclusive ethnic claims. In this context, nationalistic allegiances to Nagaland or Assam open a peculiar rhetoric about love and belonging. When mapped onto bodies, such nationalist imaginings of ethnic identities and bodies are expected to be flawless and pure. Here, the male ethnic body is constructed as a timeless bearer of purity, whereas the female body must be constantly redefined and reclaimed through marriage, ownership, reproduction, and domesticity. This concept of ethnic purity becomes the foundation for normalizing violence against women and labeling woman's body as the harbinger of impure histories and half-blood offspring in the mining towns and villages of this carbon landscape.

IMPURE LOVE

I met Lulu unexpectedly. During an interview with a coal trader in Sonari town, I paused my recorder for a moment. As I made my way to the backyard, I saw a frail woman hunched over a mountain of clothes beside the water tank. She was a part-time domestic helper, working in four or five houses washing dishes, doing laundry, sweeping the house, and picking up trash. The women from the household—the daughters-in-law of the trader—came and chatted with me about my research. When I explained that I was writing about people who lived and worked in the foothills, they pointed to Lulu and said, "She is a Naga. She also lives in Sonari town!" At that moment, I was unaware that their joy in showcasing the cosmopolitan nature of their town would lead me to encounter violent expressions of cosmopolitanism. Lulu appeared embarrassed but quickly regained her composure, and we talked a little. She asked me where was I staying, then we smiled at each other as I went inside the house to resume my interview. She was missing from my field notes as I returned to my room and noted down reflections about this mining town. I assumed that there was nothing special or important about my brief conversation with Lulu and almost forgot about her.

I was surprised when my hostess informed me that I had a guest outside waiting to see me. Lulu was sitting on the front porch. She was on her way home and wanted to talk to me. "I was not like this," Lulu began. "I was studying in high school when I fell in love and eloped with my husband," she recounted, explaining how she ended up in the foothills. Her husband's family lived in Sonari, and hence, they chose that town to settle down. Life in this foothill town was difficult for her because her husband was a drug addict. He beat her and burned her with cigarette butts. Lulu blamed it on the drugs. She said that she stayed in the marriage because of "morom lage karne" (for the sake of love). "This is not the life I imagined, but it turned out this way," she told me. She never contacted her family in Nagaland. "They must think I am dead," she said. Lulu's father had passed away, and her mother had remarried and moved to a different town. She reminisced about friends who might have graduated from college.

She saw death as the only way out of this *dukh* (difficulty). It was a violent story. I assumed she would prefer to go back to her extended family and relatives in Nagaland, but she said there was "nobody" in the hills. Neither her family nor her relatives would welcome her. The hills that she described as her *ghor* (home) were now inaccessible. Why did Lulu describe the hills as an inaccessible place? I realized that her understanding of ethnic exclusion came from her experiences in the foothills. During my fieldwork, I found that ethnic leaders in the foothills were all males, the sole authority over domestic, social, and political proceedings. Particularly in matters related to land and resources, including border disputes, they were vocal about speaking for "our state" and produced a masculine and patrilineal framework of political authority and rights.[5] These authoritative claims to represent and speak for their state were territorial and came from a process of denying women rights to property, land, and political participation. Male ethnic leaders contended that women did not have a "story" because they did not inherit land or property. Since the story of an ethnic community was situated on land whose ownership was patrilineal, women inherited the story of their husbands or fathers. Women's bodies, speech, memories, and experiences were perceived as biological organs for the male figures. Biological relationships and the family became a fertile ground for defining ethnic purity.

Ethnic purity is an important political marker, since this trait is directly connected to claims for ownership of land and resources (Malkki 1992).

Further, Malkki (1995) also describes how notions of purity played an important role in constructing ethnic identity among Hutu communities in Tanzania. The Hutu groups who lived in the camps constantly engaged in reconstructing their history of personhood and valued their suffering because it kept them pure and prepared them to return to their homeland in Burundi as the rightful natives. The camp residents saw themselves as a nation in exile, and their refusal to establish a sense of belonging and settle down in a foreign land deeply informed their cosmology of belonging. The concept of ethnic purity, according to Malkki, is eventually a moral claim that legitimizes certain practices and images of belonging.[6]

In the foothills, ethnically impure figures like Lulu were constructed as traitors. Here, the traitor was understood as a member of the community who destabilizes the ethnic mobilization for exclusive cultural identity and claims to the homeland. Adam, a tribal leader from Lulu's town, described her as a "threat" that damaged the purity of ethnic groups. Lulu's suffering remained invisible to legal and social institutions because her life was an example of all that had been forecast about the dangers of marrying an outsider. In addition, legal and political protection was reserved for the powerful: traders, businesspeople, and ethnic leaders in the foothills. When I inquired at the police station about cases of domestic violence, an officer commented, "If the husband does not beat his wife, who will beat her?"

The idea of domestic abuse being part of marital love and discipline was not limited to the police station. This belief was also rampant among those who occupied positions of power or belonged to cultural and trading alliances. Lulu's case shuttled between the legal (police station) and moral communities (tribal councils) in the town, and remains unresolved. While the police justified domestic violence as part of domestic life, political leaders and tribal councils reiterated the sanctions of their social groups and enforced purity and pollution practices. For instance, many communities marked cultural and social differences by privileging their unique history and experiences of suffering over those of others.

When I asked Adam, the tribal leader, about intermarriage, he said there were no cases he was aware of. "If a Naga man falls in love with an Assamese woman, it is a sin, and there will be a punishment. The family will fall ill or the children will die," he asserted. Inter-ethnic marriages were common, but the refusal to recognize and acknowledge such practices reproduced the fiction of ethnic and cultural purity. Such politics, according to Étienne Balibar, are about maintaining the "internal borders" of the social unit. Balibar argues that internal borders between ethnic groups and nations reflect the "problematic" politics of purity or purification and the uncertainties around which the "inside can be penetrated or adulterated by its relation with the outside" (1994: 63). What were the criteria to assess love? What was love and *not* love in the foothills?

PURE HISTORIES AND IMPURE MEMORIES

Tribal leaders like Adam wield immense power, and their opinions and decisions served as a political framework to preserve ethnic purity and culture.[7] During a conversation about Lulu's experience, he announced with an air of certainty in Nagamese, "Itu morom nohoi" (This is not love). When I inquired if he remembered any other accounts of falling in love outside the social group, he paused for a while and replied, "Except for the Dalimi story from Naginimora, we have no other story in the foothills." Even a lover of ethnic purity could not escape the story of Dalimi and Gadapani.

According to legend, Dalimi was a Naga woman married to Gadapani, an Ahom king from the eighteenth century who ruled Upper Assam in the Brahmaputra Valley. Today the legend of Dalimi and Gadapani is embraced by two groups in the foothills. The legend is popular among Ahom and Naga traders who carry out extractive economic operations like coal mining and logging in the foothills. It also has political significance in Ahom and Naga villages, which are determined to retain an alternative political history distinct from the dominant Indian nationalist ideology. The sense of communal intimacy and shared history forged by narrating this legend at Naga and Ahom social and cultural gatherings is a significant marker of attempts to create a foothill history.

What makes the legend of Dalimi and Gadapani immortal is the location of her grave. According to lore, before Dalimi died she asked the king to carry her body to her family in the hills, but instead her remains were buried in the foothills after a fierce storm disrupted the funeral convoy. Naginimora, literally "the place where the Naga queen died," sprang up around her burial site. Historically, the town of Naginimora started as a coal mining site in the early part of the twentieth century under the British administration and continued to grow as Konyak Nagas from the higher elevations came to settle there. Koloung Konyak, the oldest resident in Naginimora town, told me that he had worked in the coal mines and referred to the British coal mining operations simply as activities of the "company." Koloung described how he arrived there:

> I was born in Wakching village, but I came to Naginimora when
> I was young. I ran away from the village. They told me not to go

down, but I stole some food and came to [work] at the coal company. There were many Nepali and Bengali laborers who were working for the company. There was a huge structure here. That is where the company stood [*pointing*]. There were two wires suspended in the air that pulled the coal carts.

All the coal mining sites of the town today used to belong to the company. Those of us who worked for the company were given tea, food, and rations. We were paid fifteen paisa as wages. Rice was twenty-five paisa per kilogram. We bought at the company ration shop every Saturday. We bought our needs for the entire week on Saturdays. I was working in the coal factory and saw that people died in accidents.

But during the trouble [the Naga insurgency that started in the Naga Hills], the company was closed down. There were many people in Borjan [the present mining site]. Many laborers were working in the mines. But after the Naga movement started, rumors went around about attacks on the mines. All the laborers ran away to Assam. There was nobody to work in the mines anymore.

I was naughty, I created trouble. So, the company sent me to prison in Sibsagar jail. After I came out of jail, I started cultivating. I became a farmer. There were tigers, bears, and pythons all over this place. Then, there were huge trees here, and there was plenty of timber here. We cleared the forests here and settled down.

Koloung's memory of Naginimora demonstrates how settlements in this area were intertwined with coal mining from the beginning. The life of coal miners and the Naga insurgency illustrate the history of resource extraction and militarization here. Koloung's reference to cultivation also points to the economic and social transformation of the place. The tenor of his narration, by tracing his time in the company and then his role in clearing the forests, was similar to accounts that several residents shared with me. These stories show how people found meanings and possibilities in this place.

The legend of Dalimi and Gadapani was absent from Koloung's history of the town. Reminiscing about Naginimora as a coal mining settlement, Koloung's fading memory focused on the dense and at times fragmented mundane events of the town and the end of the British coal company that operated there. Yet it was impossible to ignore the dual importance of Naginimora as a coal mining hub and the center of the Dalimi and Gadapani legend. Several traders and seasonal workers in the foothills referred to the town as an important site of extractive economic activities such as tea, timber, sand, and pebbles. When I arrived there, I marveled at the numerous coal depots and the energy of the place as workers loaded the trucks with coal, timber, and stones. The streets were covered with coal dust, and many places in the town—homes, churchyards, school compounds, even the football field—served as makeshift coal depots. Despite the land conflicts and tensions that were routine in the tea plantations on the outskirts of this town, prospective coal traders and businesspeople from Assam and beyond traveled here. During coal season (November–April), serpentine lines of trucks carried away coal, along with timber and sand, from the upper elevations of Naginimora. Pebbles from the Dikhow River, which flowed through the town, were trucked away to Assam. Like many parts of the foothills, a section of the tea plantation lands around Naginimora town was disputed between Assam and Nagaland.

Amid the conflict, the legend of Dalimi and Gadapani remained an important part of the political life here. The story was retold in several reconciliation meetings to reiterate notions of shared history and ties of friendship and solidarity. Some years ago, the traditional council had declared Dalimi's grave an important landmark in an attempt to attract tourists, but the project fell through and the town forgot about it. Assam and Nagaland place a strong emphasis on territorial boundaries, official histories, and pure memories.

The legend is embraced, but Dalimi's grave in Naginimora is deserted. It is a decrepit concrete shed on the outskirts of town. During the summer, the road is covered by red mud. In winter, there is a layer of black dust on the grass, shrubs, and trees around Dalimi's grave due to the steady stream of coal trucks. But even though the Naginimora council forgot about it, Dalimi's grave opened a world of activities and meanings. Animals and lovers alike sought out the dilapidated shed for shelter and refuge. Children left charcoal drawings on the walls, and coal traders and miners came there for cigarette breaks. Part-real, part-legend, the stories about the grave remind the foothill residents that Dalimi rests in the foothills. Her grave site remains covered with creepers, wild ferns, red dust, and coal debris.

This legend remains part of folklore and is not recognized as an official account in the state of Nagaland. Similarly in Assam, officials, historians, and scholars dismiss this story as a legend rather than a historical account. For Nagaland and Assam alike, pure history is located away from the foothills. For instance, the higher the elevation, the purer the Naga culture and practices. Similarly, those who live in the Brahmaputra Valley believe that the essence of plains culture resides deep in the plains. Therefore, neither

Nagaland nor Assam recognizes this legend. Adherents of ethnic purity like Adam and his colleagues, who were the authoritative figures in the foothills, viewed the legend as an exceptional case. Even local historians I met during my fieldwork claimed the story of Dalimi and Gadapani was myth, due to the absence of historical sources. They agreed that Gadapani was a real historical figure in Assam but dismissed Dalimi (the Naga from the hills) as a product of Assamese literary fiction penned by the Assamese playwright Lakshminath Bezbarua. Dalimi's absence from the *Buranji*, the Ahom historical chronicles, was cited as proof that she never existed.

Historian Jayeeta Sharma (2011: 224–25) notes that citing historical sources in contemporary Assam to create national histories is complex. Official Assamese history in the twentieth century often blurred the distinction between history and fiction. Citing works of prominent Assamese intellectuals and scholars like Surya Kumar Bhuyan, Rajanikanta Bordoloi, and Lakshminath Bezbarua, Sharma describes how fictional dialogues, narratives, and characters were employed to inspire the reader. She further elaborates that "faux-historical" and "semi-fictional" accounts were integrated into Assamese history to instill readers with a sense of "sacrifice and redemption."

Sharma's insight about Gadapani's Assamese queen, Jaymati, is particularly important to understanding how the official view of Dalimi as a fictional character is part of a larger project of creating a pure history in the official nationalist imagination in Assam. In Assamese historical writing, the most popular theme is the story of King Gadapani's Assamese wife, Queen Jaymati. Legend has it that after Gadapani learned about a conspiracy in the royal court to assassinate him, Jaymati helped him escape to the hills at the cost of enduring torture and death at the hands of his enemies. Ahom and Naga villages in the foothills say that it was during Gadapani's political exile to the hills that he fell in love with Dalimi.

Sharma notes that male literary figures in Assam such as Lakshminath Bezbarua frequently borrowed from popular folklore and ballads that women sang in Upper Assam, a region contiguous with the foothills of present-day Nagaland and Assam, but failed to acknowledge the sources. She further points out that Jaymati is absent from extant chronicles of the Ahom dynasty, yet she is memorialized in contemporary Assam. March 27 is set aside as the commemoration day for Sati Jaymati, and she is frequently listed among the famous people of Assam. A local website presents her family genealogy and her life as an Ahom queen.[8]

Today, Jaymati serves as a symbol of purity, piousness, sacrifice, and loyalty for the Assamese nation. Dalimi is a foothill legend: as one ascends to the upper elevations in Nagaland or descends into the Brahmaputra Valley in Assam, this legend disappears. However, what marks the Dalimi and Gadapani story as a foothill legend is not solely its location but its richness and variation. During my fieldwork, several Naga villages offered competing accounts of their union.[9]

I came across a story from a Phom Naga coal mining village according to which Dalimi belonged to their tribe and was called Watlong. According to the Phom legend, during harvest festivals in the hills, Dalimi's family dropped red rhododendrons in the Dikhow River, which flowed from the hills of Nagaland down to the Brahmaputra Valley in Assam, to remind her of the festivity in the hills. In the neighboring villages, the Konyak Nagas made similar claims. Like the Phom, Konyak villagers claimed that Dalimi belonged to their tribe. It is significant that no single village or tribe managed to appropriate the Dalimi and Gadapani legend as a mono-ethnic narrative.

Though ethnic groups in the foothills invoke the legend of Dalimi and Gadapani to remind one another of a shared past, they strive to maintain the lines of purity. Thus, Lulu's story and the legend of Dalimi and Gadapani show how ethnic groups retain the fiction of ethnic purity to establish a social structure. At the same time, such memories are transgressed whenever necessary to start joint ventures such as coal mining operations and other land deals.

In this context, the tribal leader Adam, who referred to the Dalimi and Gadapani legend as an exceptional case and insisted on ethnic purity, also retold the legend to call for peace during reconciliation meetings between Assamese and Naga villages. However, Adam asserted that the only solution to the foothill border dispute was to maintain ethnic purity. He insisted, "Only pure ethnic groups can sit together and resolve the dispute." The solution was territorial division. The troublemakers, according to the Adam, were subjects like Lulu who gave birth to "mixed" children, and the "outsiders" who came to the foothills and settled down. He defined outsiders as "the *bagania*"—the Adivasi.

SCARRED FOR LOVE

It was common to hear stories about the lives of Adivasi. Many of them worked as daily wage laborers or sharecroppers; some of them lived on the tea plantations and did not own land, while others worked in the villages. Most of the laborers on rubber plantations, rice fields, and coal mines are Adivasi. Today, Adivasi people make up the dominant workforce on tea plantations all over Assam, including the foothills. Yet conversations about the Adivasi people often centered on gendered stereotypes of the men as lazy alcoholics and the women as promiscuous and unstable. In the foothills, these stories were detached from the oppressive structures of poverty, discrimination, and the tea plantation economy, which predominantly employs Adivasi labor. The structural violence on the tea plantation starts at the top and permeates to the domestic sphere. For instance, there was no forum to address issues of domestic abuse, because these were personal matters.

I met many leaf pickers who worked on the sprawling tea plantations in the foothills. Many tea plantations are traversed by geologists and oil exploration teams who drill on the plantation grounds looking for oil and gas. It is not unusual to witness oil rigs in the middle of tea plantations and hydrocarbon exploration on estate lands. In addition, given that the tea plantations in the foothills often operate as a de facto border between Nagaland and Assam, plantation lands are also used as coal depots during the mining season. Tea plantations and oil exploration require different scales, skills, and expertise to extract resources from the ground, but both intimately shape the social practices and experiences of the people connected to them.

At one Christian church on a tea estate, I routinely found women with bruised faces and injuries. One day, as I stopped at the church, the pastor and his wife described a recent case of domestic violence concerning a church member. The sacrificial aspect of *morom* was central to this story of domestic abuse. All the women in the church were leaf pickers. The pastor linked this act of domestic violence to the larger scheme of evil actions prompted by the devil. He lingered on the violence in the lives of women and children on the plantation. With the opening "Men drink a lot inside the plantation," he began to narrate the story of a woman I will call Savitri.

According to the pastor, Savitri's husband was a heavy drinker and disapproved of his wife's religious practices. In particular, he objected to his wife attending Sunday church service. A few weeks earlier, he had come home one evening with a hatchet and struck her on the head. With Savitri unconscious, her husband assumed that she was dead and threw her in a ditch. When her family found her, they handed over her husband to the police, but he was let off immediately. The police were not interested in filing charges, and the plantation management was also indifferent.

The most important witness in the case, according to the pastor and his wife, was the medical doctor. "You see, when the doctor saw her injury, he said the woman will not survive!" the pastor's wife recounted. When Savitri survived the injury, the doctor declared that it was impossible for the body to survive such physical trauma. Her story was recast as a miracle.[10] Savitri's survival became a story of resurrection and redemption, an expression of God's love. It became a case of distinguishing between the spiritual and the terrestrial, between the natural and the supernatural, and between dying and living. It was a story of hope and love. The pastor described Savitri's testimony to the congregation: "When she was losing consciousness in the ditch, she said she remembered God and shouted, 'Praise the Lord, halle-lujah!' At that moment, she saw a bright light and heard a voice, which told her, 'Get up.' She said the Lord gave her strength. She dragged her blood-soaked body to a paved road on the plantation and was rescued by her family, who were searching for her."

The pastor's wife emphasized that she was saved because "she cried out to the Lord to save her life." Savitri returned to her husband after the inci-dent. According to the pastor, their relationship continued to be violent. "The husband is a Christian, but he drinks a lot, whereas the wife is a true believer," he said. Explaining that she was given a new life when she took the *dhorom* (Christian faith),[11] the violent relationship was presented in terms of conflict between the profane and the sacred, between disciplining the body for God and wasting it on worldly pleasures like alcohol. Savitri's devotion and sacrifice became a point of reference for stories of violence, sacrifice, and spirituality on the plantation. Her decision to embrace the pain and violence fitted with other stories of poverty and suffering on the tea estate.

When I learned that the pastor's household did not receive any subsidy from the plantation, I asked why they stayed on. "The benefits of staying here? There are none," Easter, the pastor's wife, responded. Sometimes they took jobs as daily wage laborers in the coal mines and on the farms outside the plantation to make ends meet. Yet they endured these trials for a reason. Easter explained, "We are waiting for the door to open. It is written in the Bible, 'Ask in my name and you will receive.' It is also written, 'If you are united and ask for things in my name you will receive.' When Peter was imprisoned, the church prayed for him and he was set free. The ways of the Lord are good."

The pastor and Easter took turns emphasizing the biblical relationship between suffering and spirituality. Hardship was a way to experience the goodness and love of God. Similar to plantation workers like Savitri and her family, the pastor and his wife said they embraced their poverty and mission as a "cross." It was the cross that was the ultimate symbol of *morom*.

The pastor's family seemed deeply attached to the plantation, not because they perceived suffering as part of their journey but because they had nowhere to go. Several women in the plantation church who experienced domestic violence failed to find any forum where they might take their cases or any avenue to leave the plantation. In such situations, they sought redemption and God's love. Once within the fold of the plantation church, women sang and prayed together. How many ways are there to translate everyday experiences of violence, oppression, and loss as expressions of *morom*? Like explorers and coal traders moving through the foothills to drill down and examine the basement rock, the sediments, and the structure of the folds, I began to examine the fractured lives of gendered violence in this carbon landscape. During one of my trips to Sonari town, close to the Naga coal mines, I came across an account that captured the layered elements of *morom*. Like the shelf-slope-basinal architecture that traps fossil fuels in the form of oil, gas, and coal, *morom* trapped the social, economic, and cultural architecture of the foothill society between the feeling of loss, the desire to escape, and the yearning to belong.

EMBRACING THE FOREIGN

One day, I came across a group of teenagers exchanging remarks about the distinction between the Nagamese word *morom* and its rough English equivalent *love*. One young boy was responding to his friends, who teased him about his *morom* for a girl in the neighborhood. He said, "Hoi, hoi moi mur kukur aru mekuri-k u morom koru!" (Yes, yes, I also love my dog and cat!). But his friends reminded him, "Holiu, itu toh *I love you* ase!" (Even so, this is an *I love you*!). There was a burst of laughter and whistling, and the teasing intensified. The English phrase seemed to make the meaning of love suddenly unambiguous for the teenagers, and their exchanges become more pointed and wild. It was also striking how the teenagers seemed to connect the English word *love* with touching and sex as they dared their friend to "kiss" the girl. Until that point, the boys had played with the language of *morom*—which, as we saw above, encompasses a wide range of love and affection. *Morom* also possessed elements of fun, hope, and excitement for the teenagers, as it gave them the opportunity to improvise and create multiple interpretations of what *morom* meant. It encompassed affection for household pets, parents, as well as the girl from the neighborhood. What was the language of love in the foothills? Often young people explained romantic affairs between young people as "a case of *I love you*." The invocation of *I love you* in English appeared to convey a more definite message, almost like passing a verdict that defined the relationship between the lovers. It seemed to provide a sense of relations of intimacy unlike the messiness that *morom* brought into the picture.

Not until I visited Molong Ali, an octogenarian coal trader in Sonari town, did I realize how the English phrase *I love you* had transformed people's lives. Sonari is a vibrant coal trading town frequented by oil trucks and coal trucks. As such, most of the trading alliances between Naga landowners and Assamese coal traders are brokered in this town. Pharmacies, tailors, and hardware shops all have numerous stories about the extractive economic trade with Naga villages in the upper elevations of the foothills. When I arrived there and began to inquire about the history of the town, residents directed me to Molong Ali, one of the oldest living residents. One morning I arrived in Ali's house and asked him to share his story with me. He lived in an extended family that included several grandchildren and was well looked after by his sons and daughters-in-law. Ali sat beneath a huge wall hanging depicting the holy city of Mecca, which he had picked up during a hajj pilgrimage two years before, and began to talk about his life in this town.

I was pleasantly surprised when Ali said, "I never imagined settling in this town. I had plans to go to London." He was one of the town's most respected elders, the person who had built the first mosque in the town, and a well-regarded local historian as well. When I asked why he had decided not to go, he told me it was love that held him back. He started to tell the story of Sonari town, but it ended up as a love story about how he met his wife, Begum. They had been married for fifty years and had four children together. When the doctors in Assam diagnosed Begum with cancer, Ali took his ailing wife to the biggest cities in India seeking the best doctors to save her. They eventually returned to the foothills, and she had passed away in their house a few years earlier. In the middle of our conversation, Ali suddenly left his seat and walked over to a table. He picked up a photo of his wife and wiped away imaginary dust with his fingers. Like a car wiper blade, his creased fingers went back and forth for some time, as if caressing the portrait. "Let her also sit here," he said as he gently placed the photo frame by his side and returned to his story.

Ali was an orphan who grew up in the foothills as an apprentice in his uncle's tailoring shop during the time of the British planters, before India's independence in 1947. When he became an adult, he opened his own shop. "One day I fell ill and was running a high temperature," he said. With no one to care for him, his neighbor, an old lady, volunteered to clean his house and look after him. It was a small town, and people started wondering why Ali's shop had been closed for a week. "Even the officer from the police station asked why the tailor shop was closed," Ali said. "And your wife?" I asked him. "Wait, wait." He waved his hand and signaled me to slow down. His love story had already started, but in my hurry to catch the moment when "love" appeared on the scene, I had missed the important point: the high temperature and his bedridden, helpless body, which sparked a chain of events that transformed his life forever.

One morning, as Ali lay in bed feeling weak from the high fever, Begum visited him on her way to school. According to Ali, the doctor had visited him that morning and instructed the caregiver, an old lady who was at his bedside, to give him a bath. As his caregiver wiped his body, Begum started to help out. She wiped his body and applied oil. "Like this." He slowly picked up an imaginary bottle from the floor and pretended to pour the imaginary oil on his palm, to describe how she rubbed oil on his feet and his back. But the exchange between them was not a romantic conversation. "I reminded her to go to school. She responded that she was already late for school. Then I asked her to go back home, and she said it was late to go home," Ali said. Begum's father worked as a pharmacist on a tea plantation on the outskirts of Sonari town. When he learned that his daughter had not returned from school and was at Ali's house, he went to the police station and filed a complaint. The officer-in-charge (OC) came to Ali's residence to arrest him for kidnapping Begum.

"How could I elope with the girl? I have been bedridden with high fever," Ali informed the OC. He explained that Begum came to his house voluntarily, but there was a long interrogation session. When the OC asked Begum whether Ali had touched her inappropriately, she said no. Her father had also arrived, and there were heated verbal exchanges between them. Ali began to interrogate Begum in front of her father to prove his innocence. Ali narrated the conversation as it unfolded at this tense moment: "I asked her why she came to see me. There were so many girls in the town, and nobody came except her. I repeatedly asked her, 'Why did you come to see me? Why did you come to see me?' And she said, 'I love you,' in English."

Ali paused and looked at me. Then he brought his head forward as though whispering a secret and repeated, "She said, 'I love you.'" For Ali, the "I love you" in English conveyed with clarity what it wished to convey. It was a statement from which he could not turn back. "I told her if I wanted to get married I would already be married. I had plans to go to London and settle down. My cousin was in London and he had told me to join him there," Ali said. But now the entire town knew what she had said to him in front of the OC and her father, so no man would come and ask for her hand. Ali married her, and she became his wife.

As a young tailor, Ali was commissioned by the tea companies in the foothill town to sew the nurses' uniforms, the planter club's curtains, and the planter family's suits and dresses. He spoke about their generosity and his fascination with their lives on the tea estates. Ali described how a British planter's wife one day gave him a fish from her pond, and they became friends. She not only commissioned Ali to sew several items for her household but also gave him the idea to migrate to London and open a tailor shop. English was more than a language. It was a metaphor that captured Ali's infinite dreams and longings. It offered a clear boundary between the rulers and the subjects, and a distinction between the global language of love and a provincial language of *morom*. English was the language of hope that had captured the moment when Begum declared her love for him, and the language of hopelessness that captured his broken dream of making it to London and experiencing the English life that continued to fascinate him. But it was *morom* that captured his affectionate relationship with his uncle, his loneliness as he lay in bed sick, his insecurities growing up as an orphan, his admiration for the life of the colonial tea planters in Sonari, and the restlessness and alienation he felt in the foothill town of Sonari after his uncle passed away.

Far from migrating to London, Ali never even visited the city. He never again saw his cousins who had migrated to London. As he searched for more profitable economic avenues to support his family, he sold off his tailoring shop and never sewed clothes again. He entered the coal trade and expanded his business. He also invested money in the timber business as he became an important coal trader, and eventually rose to be an influential member of the community in the coal trading town of Sonari. Yet Ali continued to see his life as an incomplete journey. He reflected on his life in Sonari: "I never thought I would be here until the end of my life." His reflections on life and love were intertwined with nostalgia about Begum, the English planters who had been kind to him, the grand city of London he could never visit, and the unexpected position he found himself in toward the end of his life: not in London where he had desired to settle down, not with Begum who had made him stay in the foothills, and no longer a tailor. Alone and perhaps lonely and restless once again, Ali continued to wonder how he had stuck around the place he was so prepared to leave behind during his youth. It was time for his afternoon nap. I looked at the watch and realized that Ali and I had been chatting for several hours. As he concluded his life story, he used the word *morom* to express his feelings for his wife. Although Ali's life had been transformed by a declaration of love in a foreign language, toward the end of his life, he described his feelings for his wife as *morom*.

People's experiences in the foothills were deeply shaped by ideas and practices of *morom*, and were intimately connected with assertions of male ownership and gender relations. Appeals to *morom* legitimized the existing gendered politics of exclusion and violence and occupied a central place in the formulation of group identities. Everyday practices of affection illustrated the political imagination about social and moral boundaries of ethnic communities and the politics of who determined the status of subjects with rights.

Morom was neither a universal category nor a local one. Lulu's interethnic marriage condemned by the moral community, Savirti's domestic violence legitimized by the plantation management, and Molong Ali's story of love illustrate how different sites and categories—domestic, ethnic, religion, and geographical—generate accounts of sacrifice, suffering, and devotion irrespective of how disturbing and violent they are. These stories also show how people invoked the language of love to assert notions of purity, order, and meaning in the foothills. In this context, Ali's life story is not so much about how English captured the meaning of a definitive relationship at a particular moment in his life. Instead, it is about how the foreign language is connected to a larger history of extraction and shows how lives of people are deeply affected and remain entangled in the world of resource extraction. *Morom* encompasses a set of social relations that recurred on injured bodies, with unfamiliar melodies and devotional gesticulations, and moments of awkward intimacy and affection. At the same time, these relations were formed from everyday social and political ties and connections in the foothills such as the politics of purity, ethnic nationalism, mining, the plantation economy, and poverty.

CHAPTER THREE

State Loves

As I traveled through the Naga and Assamese villages in the foothills, a peculiar account of the state emerged. At times residents spoke about the state as a single sovereign entity, and at other times they invoked multiple sovereign bodies and power structures, such as the tribal councils, insurgent groups, and Indian security forces they negotiated with every day. Amid coal mining operations and the prospects of oil exploration in the foothills of Nagaland and Assam, various actors articulate their experiences of the state in the language of *morom* (love). Iterations of love to explain their relationship with the state or the absence of it allude to everyday sovereignty, power networks, and expectations against the backdrop of resource extraction in the foothills.

NO LOVE IN THE FOOTHILLS

"State love stays up in the hills. It does not filter down to the foothills. That is why our lives are miserable," Apeni said. Her reflections were common in Naga coal mining villages across the foothills. Love can be used as an analytical tool for understanding how the social worlds of affection, governance, and sovereignty are entangled here. Narratives described household participation in extractive economic activities as a necessary condition and juxtaposed this with accounts of abandonment and betrayal by the state of Assam or Nagaland, or the central Indian state, referred to simply as India. Apeni's reflection on abandonment captured the layered understandings about the state in the foothills.

During my fieldwork, many Naga residents told me, "We look after ourselves because Nagaland does not love us." According to residents like Apeni, state love was geographically demarcated for the hills and the plains. For instance, residents from her village closely compared development programs implemented by Assam and Nagaland, such as the construction of roads, schools, and medical clinics. By pointing to the absence of these infrastructures in their village, Apeni and her neighbors argued that the state of Nagaland had abandoned them.

Geographical elevation emerged as an important framework for the many foothill villages in predicting the magnitude of state love. Whether a village fell under the jurisdiction of Nagaland or Assam, residents had similar complaints. They often said that the state's love was trapped either in the hills of Nagaland or in the Brahmaputra Valley. Kohima, the state capital in the hills of Nagaland, was described as the center of power, while in Assam, Dispur was recognized as its heart of power. Naga and Assamese villagers always suspected that the rival state was more loving and affectionate. Nogen, an Assamese resident from Singibil village, explained why he felt Nagaland loved its citizens more than Assam did: "In Nagaland, the administration is very strong. They love their citizens. Only our Assam government is useless. Today, if a Naga is murdered in our village, the NSCN will come down with a jeep load of armed people and arrest the culprit!" For Nogen, the National Socialist Council of Nagaland (NSCN), a Naga insurgent group fighting for a sovereign Naga homeland in the hills, was an integral organ of the state in Nagaland. Thus, he viewed Nagaland as not merely a territorial sovereign power but a configuration of political authorities including both state and nonstate actors. Nogen's reflections on state love resonate with everyday notions of sovereignty and the presence of multiple sovereign bodies here.

However, the social construction of sovereignty is precarious. Sovereign power is not vested solely in the state but constructed and maintained by multiple groups and bodies, including the NSCN, the Naga insurgents who protect their constituencies and villages. But in crises such as natural disasters or border conflicts, state authorities from Assam and Nagaland competing for legitimacy and authority become the visible force. When I was conducting fieldwork in 2009, drought swept across Northeast India, and Assam and Nagaland were declared drought-affected areas. I frequently came across farmers who had lost their crops and distressed residents whose ponds had dried up. One day, Aben, a Naga resident, declared that the state of Nagaland had abandoned its people. Pointing toward the villages in the foothills of Assam, he said, "This year when we faced the drought and the wells dried up, the Assam government sent water tankers to the neighboring villages in the plains. The villagers were ready with their pots. The tankers came to their doorsteps with drinking water. But there is no help from the Naga-land government here."[1]

For Aben, the frequent presence of water tankers in neighboring Assamese villages (within the jurisdiction of Assam) was a manifestation of a state that cared. The visibility of the water tankers and their daily rituals of supplying water provided a framework for him to define a desirable model of the state. He said that Nagaland should learn from Assam about how to care for its citizens. Situating the well-regulated drought management system in Assam as a marker of efficiency, he went on to list a series of other state programs in Assam, such as agricultural subsidies, public schools, and hospitals, to highlight the comparatively apathetic state in Nagaland.

Apeni referred to the drought to highlight how state love failed to come down to the foothills. She said that because of Nagaland's lack of affection, her family had faced a difficult time during the drought. Their crops had failed, and she was forced to take out loans to pay her children's school fees. The family was in debt and faced a severe financial crisis. She described the state of her household and the village: "Next year, we will have to buy Assam rice. The government of Nagaland does not care for the Lower Range [the foothills]. Their help and love do not reach the Lower Range. Unlike the Middle Range and the Upper Range, where people know how to get the support and attention of the government, people in the Lower Range are in a disadvantageous position. We do not receive love from the state."

Apeni said that the state of Nagaland had abandoned them, implying the disadvantageous geographical location of the foothills, and emphasized how state love was distributed in the upper elevations of Nagaland. The Upper Range, Middle Range, and Lower Range in Nagaland are administrative categories to allocate judiciary, executive, and legislative functions. The eleven hill districts in Nagaland are divided into geographical administrative categories according to their elevation, but over the years, tribal organizations and social groups have adopted these categories to organize their activities and functions. Village councils, church administrations, school boards, and student unions allot distinct supervisory and organizational functions following the state administrative categories of Upper Range, Middle Range, and the Lower Range. This demarcation is a distinctive feature of the state in Nagaland. State officials, tribal councils, and cultural associations closely follow this spatial categorization to mobilize their power and legitimacy.

These geographical distinctions have also become a way of categorizing people's experiences and identities. Lotha Naga villagers in the Lower Range told me a story about tea that underlined how pervasive these distinctions had become. The story goes that when Lotha villagers from the Upper Range first came across tea in the *haats* (weekly markets) in the foothills, they bought the tea and carried it up to their village. Once they arrived, they boiled it until they thought it was cooked. Then they drained the water and consumed the tea leaves with rice. This story is narrated as a way of showing the ignorance of the hilltop Naga villagers, who are often out of touch in comparison to their kin groups who live in the foothill villages. Because of their easy access to the plains of Assam, the foothill villagers are considered to be connected to a wider network of markets and goods in the plains. As such, they are believed to have been the first to acquire a taste for tea, sugar, and the savories sold in the weekly markets of Assam, unlike their kin in inaccessible hilltop villages.

Another story concerned a Lotha Naga man from the hills. When he arrived in the foothills, he sat down on a tortoise thinking it was a piece of rock. When the tortoise began to move, he thought, "Ah, even rocks move here!" Such accounts illustrate how visitors to the foothills often encounter new experiences, which are not limited to human interactions. Foothill residents base their identity on their willingness to embrace new experiences. In that context, Apeni was certain that the state organs of Nagaland applied the geographical categories and cultural associations of the Upper, Middle, and Lower Ranges to allocate affection to the citizens. Her comments about the foothills as a disadvantageous location did not mean that authorities were absent. The numerous security checkpoints, the surveillance of the area, and the everyday encounters with figures of power (police, state officials, insurgents, and tribal councils) demonstrated the visibility of sovereign powers in the foothill landscape.

Everyday sovereignty is an effect of the sovereign's actions. According to Thomas Blom Hansen and Finn Stepputat (2005), sovereignty needs to be continually performed. The process of reiterating sovereignty to render it effective establishes the basic referent of the state. Sovereignty in the foothills is represented through the social relationships and encounters between authorities and the people that take place in the rigs, coal mines, tea plantations, and geological explorations. As a result, everyday sovereignty is produced through resource extraction regimes. Sovereignty creates itself in the midst of a fragmentary, uneven distribution of power and configurations of authority, exercising a form of violence that is more or less legitimate in that particular territory (Hansen and Stepputat 2005: 2–7). As foothill residents shared the virtues and models of the ideal state, a set of characteristics of sovereign powers became apparent. Here, sovereignty is produced through performances that are ritualistic, public, spectacular, mundane, and violent.

For foothill residents, Assam, Nagaland, and India constituted different forms of sovereign power. These articulations emerged from their everyday negotiations with different state entities. The creation of the states of Assam and Nagaland took place within the larger politics of reorganizing colonial provinces along linguistic lines in postcolonial India. Although the first state to be created within this arrangement was Andhra Pradesh in 1953, it was the formation of Nagaland in 1963 that highlighted the Indian state's strategy to grant political concessions and empower local elites as the legitimate representative of the Indian state in Northeast India.

A cursory look at regional newspapers and local television programs in Northeast India shows the different representations of the states, a point that people in the foothills brought up routinely. For instance, the overarching presence of Assam as the "big sister" was quite visible among the states that constitute the "Seven Sisters" of the Northeast.[2] Assam's capital, Dispur, and its twin city of Guwahati host at least four satellite television channels—all dominated by Assamese-speaking staff—which beam sensational news and views across the region throughout the day.

Guwahati is also home to a vibrant Assamese-language print media that produces news and political features for millions of Assamese-speaking citizens every day. The electronic and print media in other states, including Nagaland, are dwarfed by their counterparts in Assam. This is also true of the border issues with other states. In the foothills, the routine resource conflicts between villages within the jurisdiction of Nagaland and Assam mean that all the contentious and competing people's histories converge to represent the state of Assam or Nagaland and its citizens. This homogenization of a single identity is more visible in the foothills of Assam because residents from Assam are no longer represented by their ethnic categories of Adivasi, Bengali, or Mishing (as in the case of the Naga people), but as *Oxom-baxi* (inhabitants of Assam) who are forever under siege from ethnic groups such as the Naga, Khasi, and Arunachali people, who are defined as "encroachers." Like Nagaland, Arunachal Pradesh and Meghalaya are tribal states, and the foothills serve as their political and physical border with Assam. At times when there are heightened security alerts from the state due to land conflicts, the existence of hill states, with their special laws to prevent land transfers to nontribal persons and to regulate mobility, provide political protection from the aggrandizing effects of a combined onslaught of media and civic and administrative weight emanating from Assam. The following sections elaborate on the concept of a triadic state (Assam, Nagaland, and India).

Assam sits in the center of Northeast India, stretching in a west-east-south direction. Of the states in the region, it has the longest history of resource extraction regimes with two global commodities—Assam tea and Assam oil. It is also the most populous state, unlike the hill states that were carved out of it. Assam retains an administrative and political identity as home to tribal and nontribal communities, although this is increasingly challenged by political mobilization demanding autonomous councils and separate statehood for different ethnic groups.[3] The creation in 2003 of the Bodo Territorial Autonomous Districts, governed by the Bodo Territorial Council, and demands by tribes in autonomous districts like Karbi Anglong and Dima Hasao (earlier known as North Cachar Hills) have resulted in some power devolving to these districts.

In quantitative terms, factories and industries established in Assam outnumber those in neighboring hill states, and the state is recognized as an economic hub. Under such circumstances, foothill border disputes between Assam and neighboring hill states, including Nagaland, often result in economic blockades, accentuating the character of Assam as the state that controls the economy and movement of goods and traffic for the entire region. Popularly referred to as the Gateway of Northeast India, Assam possesses and regulates the movement of people and commodities in the region, and at times uses its economic strength and geographical location in the valley to intimidate its neighbors in the hill states.

Political demands for tribal autonomous regions from different ethnic groups within Assam continue to be contentious, couched in a language of ethnicity that alienates other communities (Barbora 2005). Every demand for an ethnic homeland within the existing state of Assam evokes opposition from a constituency that feels left out of the ethnic mobilization. As a consequence, the state has become a repository of the different permutations and combinations of ethnic politics in Northeast India. On the one hand, its multiethnic composition is undermined by tensions created by differences in language, religion, and location, in which non-Assamese-speaking communities (especially indigenous tribes) have emerged with radical demands for separation and devolution of power (Baruah 1999). Nontribal communities, such as the Adivasi, Nepali, and Bengali-speaking Muslims, are quick to resist demands for any change in the status quo.

This partly has to do with the nature of ethnic politics in Assam and partly with the way electoral politics and constituencies are organized. While most ethnic and tribal communities expect to mobilize along ethnic lines for greater territorial autonomy over areas where they have a numerical superiority, it is impossible to think of dispersed groups like Adivasi, Nepali, and Bengali-speaking Muslims having similar political aspirations. These dispersed groups prefer political structures that ignore ethnic identity and control over resources, favoring the tested political principle of majoritarian rule.[4]

Before 1947, Assam was a province in colonial India sequestered by partially administered and excluded areas in the hills. Sanjib Baruah describes the strategic colonial goals of securing the Indian heartland by creating frontier areas during the latter half of the nineteenth century. Assam and its strategic hinterland in the hills became known as the North East Frontier Province and shared a colonial administrative touch similar to the North Western Frontier Province in the northwest (Khyber Pakhtunkhwa, as it is called now). Even during colonial times, therefore, the province of Assam was governed under different rules for the hills and valleys.[5]

The process of creating separate federal units from the province of Assam began with the creation of Nagaland in 1963, and continued with the creation of Meghalaya (comprising the Khasi, Jaintia, and Garo hill districts) in 1973, and Mizoram (comprising the Lushai hills) in 1987. These federal units partly reflect a central government strategy to create manageable linguistic and ethnic states that would be forever dependent on central fiscal grants and partly reflect the intransigence of Assamese upper castes that wanted to foist a single language and culture upon the people of the hills (Baruah 2005; Chaube 2012).

In the nineteenth and twentieth centuries, the colonial administration in Assam encouraged a large number of Bengali-speaking cultivators to immigrate from the densely populated districts of the neighboring province of Bengal. This was done to resettle displaced peasants and to increase revenues from land, since the native peasantry was not used to paying taxes for settled agriculture (Baruah 2001). This demographic change remained a recurring issue in political mobilization in Assam, with the Assamese-speaking political elite accepting the referendum in the Bengali-Muslim district of Sylhet that made it part of Pakistan in 1947. The Assamese political elite feared losing power to a numerically dominant Bengali-speaking community. However, the partition of the British Empire had a particularly disrupting effect on politics in the Northeast, especially in the densely populated valleys.

In 1979, students in Assam and Assamese nationalist fronts began a civil disobedience movement to demand the expulsion and disenfranchisement of "foreigners"—mainly Bengali-speaking Muslims—from the state of

Assam. In doing so, they revived an early twentieth-century fear prevalent among Assamese nationalists that unchecked migration into Assam would result in the Assamese and other indigenous communities losing political and economic power. The All Assam Students Union and Assam Gana Sangram Parishad (Assam Peoples Struggle Front) started the movement, and their demands were directed to the central government in India. The Indian government was placed in a precarious position, for to accede to the movement's demands to treat all Bengali-speaking persons as illegal immigrants would deny Bengali Hindus a refuge from religious persecution in Bangladesh (erstwhile East Pakistan).

Furthermore, to give in to demands to evict Muslim Bengalis would also mean flouting the secular principles of the constitution and inviting diplomatic conflicts with Bangladesh, which denied that its citizens were illegally entering India.[6] In the ensuing impasse, thousands of people were killed—both by security personnel and nonstate armed groups—until an accord was signed between the movement's leaders and the prime minister of India in 1985. In the meantime, Assam's political mobilization had taken a radical turn with a section of the youth turning away from nonviolent parliamentary politics and espousing armed rebellion (Hazarika 1994).

With the formation of the United Liberation Front of Assam (ULFA) in 1979 and the organization's growth in the decades that followed, Assam became wracked by violence and uncertainty. By the late 1980s, radical youth in most indigenous communities, including the Bodos, Karbis, Dimasas, and Adivasi, formed armed groups to assert their communitarian rights over land and natural resources. Throughout the 1990s and early 2000s, these groups and the ULFA disrupted political and economic life in the state. The government's response—a combination of brute strength and pecuniary incentives—succeeded in driving a wedge between the armed groups and bringing many of them to talks by offering them the opportunity to participate in India's economic growth and development (Barbora 2005).

NAGALAND: CULTURAL STATE

Nagaland functions as a cultural state. The official website of the government of India declares Nagaland to be "the land of festivals,"[7] and the official webpage of the state government describes the Naga people as "lovers of fun and frolic" and life in Nagaland as "one long festival."[8] However, the conflict between Naga armed groups and the government of India, recognized as the longest insurgency in South Asia, tells a different story.

Soon after India's independence in 1947, an armed uprising led to formation of the Naga National Council, an armed group fighting for a Naga homeland. On December 1, 1963, Nagaland became the sixteenth state of the Indian Union. Thus, the formation of Nagaland took place in the midst of an armed confrontation between the Indian army and Naga armed groups. The formation of the state was the outcome of an agreement between the politically moderate Naga Peoples Convention (NPC) and the government of India. The NPC members who negotiated the formation of Nagaland were erstwhile members of the Naga National Council.

An official document known as the Sixteen-Point Agreement laid the framework for the formation of Nagaland in 1963. The legislative, executive, and judiciary functions of the new state were enshrined in the constitution with the Thirteenth Amendment Act of 1962.[9] In recognizing the special provisions for the federal unit, Article 371A of the Indian Constitution guaranteed certain rights to the Naga people residing in the state that were different from existing provisions for other communities and regions in post-Independence India. The clauses in Article 371A broadly followed the concessions noted in the Sixteen-Point Agreement.[10]

The formation of Nagaland also coincided with the first Indo-Naga cease-fire, which was initiated in January 1963. However, Indo-Naga negotiations fell apart in 1975, thirteen years after the state of Nagaland was created. With the signing of the Shillong Accord in 1975, a section of the Naga armed group from the Naga National Council (NNC) unconditionally accepted the Constitution of India and acceded sovereignty to the government of India.

In the aftermath of the Shillong Accord of 1975, there was an escalation of violence among various Naga political groups and armed wings. In 1980, a section of the NNC that had opposed the Shillong Accord established a separate armed group known as the National Socialist Council of Nagaland (NSCN). In 1988, the NSCN was further divided into the NSCN (Isak-Muivah) and the NSCN (Khaplang), each faction named after its leader/s.[11]

The NSCN(IM) and the government of India entered into the second Indo-Naga cease-fire agreement in 1997.[12] Since the Indo-Naga cease-fire of 1997, Naga insurgent groups have undergone further divisions. The number of casualties from violence between various factions of the Naga armed groups has escalated during the period of the cease-fire even as confrontations between the Indian security forces and the Naga insurgents continue to take place.[13] This political history underlines how the multiple Naga armed groups, the government of Nagaland, and the Indian state (represented by the security forces involved in counterinsurgency operations) highlight different forms of legitimacy and power in Nagaland today. Both organs—Nagaland and Naga armed groups—enjoy significant sovereign status in Nagaland while the Indian security forces are governed and authorized with extraconstitutional powers under the Armed Forces Special Powers Act (1958).

Since Naga armed groups entered into a cease-fire agreement with the government of India in 1997, the state of Nagaland has been promoted as a tourist destination and a prospective state for development programs to promote peace and harmony. Nagaland enjoys a special constitutional provision. Article 371A of the Indian Constitution grants Nagas in Nagaland autonomy with rights over community land and natural resources, including the right to govern themselves though customary law. Naga villages in the foothills invoke this provision to legitimize coal mining operations and hope to use it to negotiate oil exploration and obtain royalties in the near future. In addition, Naga customary law allows Naga trading families to claim ownership over their land and natural resources, including coal and timber.

Because the recognition of community culture, customs, and ownership of land and natural resources is guaranteed by the constitution of India, extractive resource operations like coal mining and logging are often defined as a cultural practice. Yet there is serious conflict over who represents the people in Nagaland. Naga insurgents, village councils, and government organs all compete to represent the Naga people because representation, political identity, and Naga culture are tied to control of land and natural resources in the hills.

Investment Opportunities in Mineral Sector in Nagaland, a brochure published by the Directorate of Geology and Mining of Nagaland that focuses on the possibilities of a carbon future, states, "Nagaland, a tiny hilly and tribal state, nestled in the northeast part of India, is endowed with substantial mineral resources, having potential for economic growth development. These are offered for exploration and exploitation to the prospective investors, both domestic and foreign." Yet such hopes are not shared by all. During my fieldwork, a cultivator from a Naga village in the foothills described the ongoing land deals in Nagaland: "If you like one mountain but there are a certain number of landowners there, you can do a Memorandum of Understanding and buy the entire mountain. See, in another fifty years the poor will not survive in Nagaland. All the rich people will buy off the land, and we will become poorer and worse than the beggars in the plains [Assam]."

Beneath the facade of a cultural state based on an exclusive politics of belonging, ethnic identity, and kinship ties, there is growing anxiety about the privatization of land and resources in the state. The customary protection enshrined in Article 371A of the Indian Constitution remains ambiguous about who represents the people—the state or the tribal councils—and does not define what constitutes Naga customs and culture. This opaqueness has led to a return to patriarchal and reactionary practices regarding gender relations and management of natural resources, especially in the Naga coal mining villages where various authorities and actors—tribal chiefs, politicians, insurgents, police, Assamese traders, and landowners—together define tribal rights and authority. By combining a masculine and exclusive rhetoric of Naga culture, power, and inheritance, the extractive resource operations in Naga coal mining villages are described as cultural practices. During negotiations between Naga landowners and non-Naga traders, customary law is invoked in business deals.

Thus, the overlapping sovereign powers in Nagaland, the constitutional caveats and restrictions on oil exploration (by Naga insurgents in the past), and the dispute over who owns the land—the Naga people or the government of Nagaland—mean that any changes in land policy are bound to be far-reaching and irreversible in the near future. Since the Indo-Naga Agreement of 1997, debates over natural resources and the wealth-producing capacity of the land have brought the question of representation to the foreground. As various groups—some armed, such as the Naga insurgents, and others elected, such as representatives in Naga tribal councils, student bodies, cultural associations, and the state legislative assembly—vie for the right to represent the Naga past and future, this dispute focuses primarily on control of land with oil and coal deposits in the foothills of Nagaland.[14]

INDIA: MILITARY STATE

People's accounts about the triadic states in the foothills allow us to comprehend everyday understandings of multiple sovereignty, but they also invoke the regional experiences of the modern state. As in Africa, where the postcolonial state is an inheritor of colonial power, authority, and violence (Mbembe 2001), in Northeast India the relationship between the Indian state and its citizens has been violent and militarized (Samaddar 2001; Baruah 1999). In the foothills, Indian security forces are referred to as employees of "India" or "Delhi," and extraconstitutional regulations such as the Disturbed Areas Act (1955) and the Armed Forces Special Powers Act (1958) that establish a militaristic system of governance across the foothills are called "India's law."

Security camps, townships, watchtowers, checkpoints, and surveillance across the landscape offer a vignette of everyday militarization here. The segregation of security and oil townships from civilian villages and towns produces a peculiar lived understanding of militarization. For the oil technicians and the security forces posted here, the state of exception lies out there, in the villages and towns that surround their barricaded townships, oil and gas gathering stations, and rigs. These villages and towns are considered violent and dangerous places, while for the people living within, the gated communities represent structures of violence and the Indian state.

Assam became a province after the transfer of power from Britain to newly independent India in 1947. Today, the state of Assam has become an economic powerhouse, as its name is attached to commodities like Assam tea and Assam oil. Yet revenues from these resources are unevenly distributed, with a substantial portion taken out of state. Nonetheless, neighboring hill states such as Nagaland regard Assam as a model—so much so that traditional bodies in Nagaland negotiating for resuming oil exploration have demanded the right to use the names Naga oil and Naga gas. Such ethnicization of hydrocarbon does not erase the powerful presence of the Indian state as a military force in Northeast India.

But from a geopolitical and security standpoint, the region is categorized as a politically sensitive zone and requires extraordinary surveillance and regulation. Given that the foothills are both a troubled area and a prospective hydrocarbon zone, it is impossible to ignore the presence of multiple state actors (officials, ethnic bodies, and security forces) and nonstate authorities (insurgents) here. Across the foothills, curfews are imposed several weeks ahead of India's Independence Day (August 15) and Republic Day (January 26) celebrations for security purposes. Security is heightened to monitor the traffic of people and goods across the region.

In addition, the Central Reserve Police Force (CRPF) also monitors the foothill villages. In particular, along the Disputed Area Belt the CRPF checkpoints regulate the movement of grains, vegetables, and construction materials like sand and bricks because construction of new structure is prohibited. Given the nature of the border conflict, where the state of Nagaland and Assam have overlapping claims on the foothills, the CRPF serves as a neutral force that reports directly to New Delhi.

This intimately informs people's local knowledge about the multiple sovereign authorities. Mentions of the state—Nagaland, Assam, and India—in the foothills alternate between references to distinct federal units of Nagaland and Assam, and to the nation-state of India, usually via references to New Delhi.[15] Foothill people's references to the multiple states in their lives reveal everyday state practices and the ways people negotiate these multiple sovereign presences.

While Nagaland has cultural claims and Assam functions as an economic state, it is India as the security state that emerges as the most powerful force across the foothills. The imposition of extraconstitutional regulations, such as the Armed Forces Special Powers Act (1958) and the Disturbed Areas Act (1955), gives the armed forces extraordinary powers. Yet the deeply segregated worlds of Indian security forces, residents inside the oil townships, experts working in the rigs, and people in the villages are framed by anxiety, suspicion, and fear. Their points of encounter are often under the surveillance of watchtowers or checkpoints. This deeply militarized separation of the extractive world and the social world of the foothills produces a peculiar understanding of the state of exception. The danger is always something that lies out there, in that other group, and not within us.

On May 6, 2015, the *New Indian Express* reported that the states of Assam and Nagaland had deployed additional security forces on the foothill border after a land conflict between two villages turned violent, leading to destruction of property. According to the report, while security forces of the rival states built bunkers, "as part of the plan to brace for any eventuality," a representative of the Indian Civil Service, the deputy commissioner, surprised at the turn of events, suggested that the security forces of Assam and Nagaland at the foothill border should refrain from attacking each other and maintain peace.[16] Once again, the CRPF was called in as a neutral force to control the mobs in the towns and villages along the foothill border.

In the context of locating the concept of everyday sovereignty in the foothills, the work of Edmund Leach (1954) and James Scott (2009) offers important historical and political background. Leach described two kinds of polity in Burma and highlighted the relationship between them, while the third entity that shaped political discourse—the colonial state—looms in the background. Similarly, Scott does not deny the rise of the modern state, but he framed his work within a hill and valley paradigm that excluded in-between spaces like the foothills. The experiences of residents from the foothills of Assam and Nagaland offer a framework to understand state sovereignty and authority as a triadic force of power and authority seen in the foothills.

This logic is dangerous and frames the Indian state as a militarizing entity when the Indian armed forces, perceived as the institutions and structures of the Indian state—or "India," as they say in the foothills—barricade themselves inside gated communities and camps, only emerging to impose order or regulate oil exploration and the movement of tea or to enable the coal trucks to pass peacefully. Especially in the foothills of Assam, Indian security forces like the Central Industrial Security Forces (CISF) perceive tea, oil, and gas, commodities extracted from the earth, as neutral, peaceful items required for the growth and development of the country, while the citizens who live on the land where extraction takes place are viewed as dangerous and untrustworthy. Such practices further reify the perception of the Indian state as militaristic.

Unlike Assam (as the economic state) or India (as the military state), the rise of Nagaland as a cultural state is founded on the principle of a tribal state where Naga people will recognize the state as the rightful head and overseer of customary practices and history, including its land and resources. This means bestowing a sense of tribal affection where the idealization of a hill identity as exclusive and pristine is routinely staged to project the hill state as the sole authority.

HILL LOVE

Emphasizing the history of affection between hilltop Naga villages and the state in Nagaland, the chief minister addressed a large gathering in Mon town: "Keeper of Naga culture and tradition, king of Chi village, you have kept the identity of the Nagas and I am happy. I knew his father and now I have met his son and we have renewed the relationship, and that is why I have come here. Among the Nagas, the Konyak people have retained the culture of the Nagas since the time of British. Nagas are all proud of the Konyaks. As a token of our appreciation and love, we have distributed gifts to all the village elders."[17]

The gifts for the village from the state of Nagaland consisted of the following items:

1. Fifty bags of rice
2. One hundred and ninety plastic rolls
3. One steel cupboard
4. Five woolen blankets
5. Two hundred and twenty synthetic blankets
6. Two hundred cups
7. Two hundred and fifty plates
8. Twenty bundles of corrugated galvanized iron (CGI) sheets

During this 2010 official visit to Mon district, the home of the Konyak Naga tribes, the chief minister of Nagaland, Neiphiu Rio, framed his vision of state love and development. This district has the most coal mines in Nagaland. The increasing deforestation and contamination of water and soil due

A state function in Chi village (Nagaland)

to coal mining across the Naga villages in the foothills have led to a series of debates about unscientific mining activities in the state. However, similar to neighboring districts, coal mining communities in Mon district have resisted the state's call to ban coal mining activities.

Konyak tribal areas have been central to the history of coal in Nagaland. The British administration opened the first coal mines in Konyak tribal areas of the foothills like Naginimora in the early twentieth century. From the beginning, British administrators negotiated directly with Konyak tribal bodies in matters related to coal extraction (Saikia 2011). Throughout the coal mining areas of Mon district, there were conversations about routine consultations between the Konyak cultural and traditional associations and the government of Nagaland over monitoring the mining activities. The Konyak tribes have resisted the state's attempts to ban coal mining and condemned state directives to ban coal mining as jeopardizing their livelihood (McDuie-Ra and Kikon 2016).

Against the backdrop of a contentious debate over coal mining, the arrival of the head of the state, the chief minister (CM), became an important event. The CM announced, "I have adopted this village. This is my village, so whatever you need I will give you. But you will have to be my partner. It requires a man and woman to make a baby. I cannot make a baby on my own. If Chi village is not my partner, then there can be no baby. When a man and a woman sleep together, they have a baby. In our case, our baby is development." Then he began to describe the gift he had brought from Kohima, the state capital, for the Konyak chief. It was a sports utility vehicle, the latest model of the Mahindra Scorpio jeep. The official purpose of the CM's visit was to lay the foundation stone for the Naga museum in the village, but this was seen as an opportunity to start conversations about the ongoing coal mining activities in the district. Nagaland's increasing anxiety about losing control over mineral resources, including coal and oil, was apparent.

Given the long history of coal politics driving a wedge between the tribal communities and the Nagaland government, the CM took the time to spell out how the tribal state cared for its people and elaborated a plan to develop Konyak villages like Chi as tourist attractions. This involved building a Naga museum in the district. Announcing that the objective of the museum was to create a space where "we" would store and preserve all "our" Naga artifacts and showcase them to the world, the CM taught the Konyak audience about the significance of the artifacts that would be displayed in the museum. He used the Nagamese phrase "Itu toh ineka ineka ase aru outoh toh ineka ineka ase" (This is this and this, and that is that and that). The CM was using a phrase in Nagamese to indicate how the Konyak people should equip themselves with knowledge such as dates, events, and histories about every material object displayed in the museum. In that sense, the Nagamese phrase employs a fill-in-the-blanks format where the Konyak would add the correct details about their culture and history. The CM's speech in Nagamese was pedagogic and condescending.

"Unless the people are close to the Government, democracy will be bereft of its spirit and meaning," Mr. Rio said at the Tenth North East Region Commonwealth Parliamentary Association Conference in 2007. Underlining how the Indian parliamentary system failed to adopt the "time-tested traditions, customs, and aspirations of the people," Rio proposed that the participants embrace "parliamentary diplomacy" to govern difficult regions like Northeast India. He emphasized that state legislators who carried the mandate of the people in Northeast India needed to learn to be leaders and fulfill the "dreams and aspirations" of the people they represented.[18] Employing the language of kin love and care, the CM sought to integrate Naga customs, culture, and traditions in his model for regulating and controlling the coal mining activities that had been administered by landowners, traders, and traditional bodies.

In this regard, the museum in Chi village became a symbol of state love and affection. The CM's address to the Konyak people as "keepers of Naga culture and tradition" showed how state authorities viewed the hill state as a place of purity and authentic Naga culture. The CM's speech also seemed to engage the notion of a homogenous Naga culture to define the tribal state in Nagaland. His reference to the Konyak people as keepers of Naga history and culture became clear as he gave his interpretation of Naga history:

> The British came and saw that Nagas are [a] different race [from] the plains people. The Nagas should be protected; otherwise they will suffer. The British people in the British parliament in 1873 passed a bill. That is why we have the ILP [Inner Line Permit]. We have protection because of that. See, the plains people have lots of money and they can come up to the hills and buy up all our land. Moreover, Hindus, Muslims, Buddhists—there are so many religions down in the plains; they could come up and break our tradition and rule over us. Inner Line Permit (ILP) is a protection; that is why we are thankful to the British. After the British left, when we were under the rule of India, there were restrictions for foreigners, so we got Restricted Area Permit (RAP) after independence. After that, we signed 16 Point Agreement, we got special provision under the Indian Constitution. We got special protection for our culture, tradition, and practices. India said, "Use all these special ways to govern yourselves." We got Article 371A! All this means outsiders recognize Nagas but we do not recognize ourselves. Why is that?

The speech described how Nagaland was founded on notions of spatial and cultural purity because the hills were always protected from the chaos and impurity of the plains. He listed colonial regulations and agreements such as the Inner Line Permit and the Restricted Area Permit, which regulated the flow of travelers to Nagaland, and the Sixteen-Point Agreement, which led to the formation of Nagaland, as well as Article 371A, which grants special resource and land ownership rights to the Nagas in Nagaland. According to the CM, the colonial administrators had been benevolent toward the Naga people at least in that they protected the hills from the suffering and chaos of the plains, and he seamlessly connected the colonial regulations with political and constitutional developments in postcolonial India. A framework emerged for the foundation of Nagaland as a cultural state. Yet, foothill residents from Nagaland had different experiences.

Ben, a retired government employee from a Naga village in the foothills, described the inefficient and corrupt system at work in Nagaland: "The Nagaland state functions as a band of thieves." For example, a couple of years earlier, around thirty households submitted an application for government relief after their *jhum* fields were destroyed in a forest fire. Ben said that "New Delhi" sent money but the authorities in Nagaland did not distribute it to the victims. Such relief money and similar development packets were all "eaten away along the way." Ben and his friends explained how officials stationed in Kohima, the capital of Nagaland, ruled the state. The rulers in Kohima kept for themselves much of the money for developmental projects and relief efforts. The money that managed to trickle down to the district headquarters was distributed to those in the good graces of the state actors—politicians, bureaucrats, and households with a strong connection to insurgent groups—while others were left out.

"This is what I will say: it is the state government who is the enemy; it is this government who plays games with us so that we cannot survive. Government does not love us. The Nagaland government is a government to oppress all those who are not connected to them. All the farmers suffer, and there is no way we can benefit," Ben concluded. Accounts of state corruption, along with conversations about feeling abandoned, frequently came up in these villages.

One day, I sat with a retired schoolteacher in a Naga village and listened to stories about corruption. I expected routine stories that I had heard in the neighboring villages and nodded as the schoolteacher said, "The village chairman and secretary eat up the money. There has been no initiative even to distribute the NREGA job cards, but the Nagaland government does not check all this corruption!" Suddenly his wife, who was listening to our conversation, whispered, "Hey! Do not bring up the NREGA name!" Programs like the National Rural Employment Guarantee Act of 2005 (NREGA)[19] and local Village Development Board development projects, such as cleaning roads and drains, were implemented in the villages. Concerned about the flash of fear that took over the wife, I asked, "What is it with the NREGA name?" It appeared that the finance secretary of the village had recently run off with all the NREGA project money, around 7 lacs (US$15,000), and joined a Naga insurgent group. This was considered a shameful and unfortunate incident.

But similar stories about residents taking off with state funds and property continued to surface. Some years ago, there had been an acute scarcity of water in the village where the NREGA money scandal had taken place. Ari, the village headman, blamed the residents for the water crisis: "The pump installed to bring water to the village has been stolen twice. That is why the village water tank is empty and there is no water. The first time someone stole it to set up a water supply while constructing his house, and the second time residents stole it to irrigate their farms and bring water to their fisheries." When state officials learned about the thefts, they stopped development work in the village. The department workers took the remaining water pipes to a different location.

Other state property faced a similar fate. People stole or destroyed various state installations—solar panels, electrical wires, metal water pipes, water pumps, and telephone cables. Everything was removed, sold, exchanged, or converted for other uses. Arjun Appadurai (1986: 26) notes that the diversion of commodities from specified paths indicates either creativity or crisis driven by aesthetic or economic forces. These incidents in the foothill villages reflected both. Out of the crisis of a dysfunctional and militarized structure, residents found creative ways to take care of themselves. Government utilities in the villages were constructed but seldom maintained. Damaged transformers took months to be fixed and were covered with creepers and wild grass, broken water pipes protruded from the ground, and dilapidated government offices, schools, and medical centers became shelters for cattle.

The absence of state love was reflected in the dilapidated infrastructure of the villages. Naga residents in the foothills told me, "Look around this village. Do you think [the] state loves us?" From the missing telephone lines, broken water pipes, and department offices in ruins, what emerged was the absence of a caring state. This feeling of abandonment spurred the Naga villagers to contemplate what constituted a loving and benevolent state.

THE HEARTS OF THE PUBLIC

Naga villagers found a model of the perfect state elsewhere in the foothills. "Of course, people in Assam are better off!" Jami said. In 1995, when there was a land conflict on the outskirts of his village, fifty-four cars from Assam arrived carrying important administrators, army, police, and border forces. But there was no sign of any state officials from Nagaland. "See, the Assam government provides everything to their people, from tin sheets to agricultural projects and roads. But our Nagaland government does not care. They do not give us anything." Then Jami quizzed me, "Do you think the state loves us? Do you think the state cares for us? Have you seen the condition of our roads?" The last question referred to the crumbling infrastructure in Nagaland. Contextualizing good roads as symbols of state love, Jami said, "One can gauge the economic well-being of a village simply by observing the condition of the roads." Several other foothill residents also said that good roads reflected the status of development programs and relief packages. Naga villages made this connection by comparing the muddy and broken roads in their villages with the four-lane highways in Assam. "In Assam, the state loves its citizens. Look at their roads. It says everything," Jami said.

Misuse of department funds and corruption is routine in Nagaland. The 2011 *Kohima District Human Development Report* describes the existing infrastructure and connectivity in Nagaland as "dismal."[20] Of the total road network of 13,371.45 kilometers, more than 50 percent was "unsurfaced." Of that, 90 percent of the areas without proper road connectivity were rural. The report stated, "90 percent of the villages have only fair-weather approach roads" (Kathipri and Kechu 2011: 137). Given that 75 percent of the population in Nagaland live in rural areas and are primarily subsistence cultivators, good roads are important to transform the state from an agriculturally driven economy into an industrialized society (Kathipri and Kechu 2011: 137–38).

One day, the seed farm manager at Nagaland Seed Farm in Merapani commented that there were no good roads in the foothills, and this was the reason why the state of Nagaland "did not come down." He added, "If only the state of Nagaland knew the hearts of the public in the Lower Range, everyone would benefit." When I asked what he meant by the "hearts of the public," he said, "The hearts of the public are attached to their land."

"The Naga government does not understand agriculture. That is why it is difficult [for the state] to love the public." The seed farm manager directed daily activities and functions in the foothills. Established in 1969 as a seed production and seed multiplication center, Nagaland Seed Farm sat on 542 acres of land. Like many places on the foothill border, the seed farm was also one of the sites where the rival states of Assam and Nagaland had overlapping claims. A Central Reserve Police Force camp was set up adjacent to the seed farm to monitor the foothill border between Assam and Nagaland. The security guards posted there watched over the seed farm throughout the year.

This farm operated as a public distribution center to provide rice and supplied seeds to the government of Nagaland for distribution among cultivators who practiced wet rice cultivation in the state. It was also an experimental station for high-yield crops where local scientists from the state applied rice intensification technology, an agricultural method to increase production. There is sparse documentation on the farm, but I came across a 1976 official evaluation of the seed farm conducted by the state of Nagaland. This evaluation spelled out the future of agricultural transformation in Nagaland and provided a list of recommendations:

1. Improved farming techniques should be adopted to achieve full utilisation of existing production capacity [about 51 percent of production capacity was unutilized]. The Farm's production should be raised by extending cultivation to the 95 hectares that had been reclaimed, but not cultivated.
2. For diversification of production, it is necessary to increase production of crops other than paddy.
3. A Seeds Advisory Committee should be set up to properly handle the Farm's production plans and review the progress of their implementation.
4. The Farm should try to raise its revenue from the sale of produce and supplement its income by developing orchards on some of its cultivable land lying fallow.
5. A specific study may be conducted on the possibilities of developing river lift irrigation and evolving [a] suitable system of crop rotation, with adequate stress on multiple cropping.
6. For seed treatment, the Directorate of Agriculture should provide seed dressing drums and chemicals along with facilities for practical demonstrations. Two storage sheds should be constructed in order to have adequate storage facilities at the Seed Farm. The Farm should be provided with electrification, which would facilitate the use of power tillers, winnows, threshers, pumping-sets, etc. at much lower costs.
7. The Farm should switch over to metric system of weights and measures, since the present system of weighing crops in measures of "tins" is very crude.

(Government of Nagaland 1976)

The advisory recommendations and the steps to adopt scientific measurement units and agricultural practices were important, but managing the seed farm was challenging. Because of the border dispute, the court had prohibited construction of structures, including fences and gates, in many parts of the foothills. As a result, cattle from neighboring villages grazed over large tracts of state farmland. In addition, Oil and Natural Gas Corporation teams conducted oil and gas exploration in fields adjacent to the seed farm. The routine encounters between the Naga farmers and geological experts, as well as the armed security forces who protected the hydrocarbon activities, resulted in frequent frisking and police checks on employees.

According to the seed farm manager, due to the surveillance and militarization of the foothills, residents in the Naga and Assamese villages adjoining the farm adopted different ways to resolve conflicts in their respective villages and neighborhoods. For instance, people from Assam immediately informed their state agencies and legally settled their claims through revenue offices and other official channels in Assam. By contrast, Naga villages appealed to customary law and village councils during conflicts and disputes. Naga villages had little to do with the state of Nagaland when settling land and property matters. This was primarily due to the strong commitment of the Naga coal mining villages to keep the state from controlling the coal mines and their community land.

These differences, according to the seed farm manager, also reflected how the respective states responded and connected with their citizens in the foothills. He said that the All India Radio station in Assam seemed to exemplify the affection of the state. "When I turn on the radio . . . I do not speak Assamese fluently, but the radio station will be guiding the farmers in Assam as to what crops they should sow in this season," he said, summarizing the wealth of information: "The radio directs the farmers to collect the seeds from specific centers. If the center is in Titabor town, then the radio will give directions to the farmers to go to the center and collect it. See, the Assamese farmers know more about tomato cultivation than the Naga farmers in Nagaland. The Assamese farmers know which variety, what pesticides, and what fertilizers to use—they know better. They turn on the radio and cultivate. They know about the pesticides better than the hill farmers!"

Every time he listened to Assamese agricultural programs, he was impressed with how farmers in Assam were more informed about agricultural practices than the cultivators in Nagaland. In contrast, the state of Nagaland did not focus on agriculture and land issues. To illustrate his point, he summarized the news on the All India Radio station in Kohima:

> Even yesterday, when I listened to the radio, people in the plains are talking so much about different types of plants, varieties in the fields and the crops. But for us, on our radio stations there were programs about how successful this or that jubilee was in that village, who was the main speaker, what did he say. Then other news was about cultural dances and songs. I thought, "Suits you fine!" They will not tell us any other thing. But the

Assamese radio will be telling the farmers even about the pests in the fields and the kinds of crops that attract certain pests and how they should be treated, what pesticide to use. The radio will even tell the farmers which shops sell what kinds of pesticides.

His ability to understand multiple languages spoken in the foothills made him and other foothill residents from Nagaland long for a state that showed affection to its citizens. Eventually the seed farm manager confided that his professional journey had been full of suffering. The state of Nagaland failed to recognize him as a dedicated employee who had devoted his life to demonstrating that agriculture was the way to connect Naga residents with the government. The seed farm manger's desolation was profound. Yet, it was part of the everyday experience of living in the foothills.

CONCLUSION

The relationship between the residents in the foothills and the state emerged as a tale of suffering and anguish. The tales of abandonment and lack of state affection were tied to the belief that state love was either at the higher elevation of the hills (in Nagaland) or in the valley (in Assam). In addition, the militarization of the foothills resulting from the presence of the security forces gave rise to a distinct foothill experience of seeing the state as multiple sovereign authorities. For residents in the foothills, Assam, Nagaland, and India were distinct sovereign powers in addition to the nonstate actors who operated here. In particular, the state of Assam as an economic power, Nagaland as a cultural state, and the Indian state as a military authority drew out the different constitutional, extraconstitutional, and historical aspects of the formation of these powers. Given the coal mining activities and the state's claim as a tribal entity to govern and control natural resources including land, the emergence of a cultural state in Nagaland could be seen as a drive to consolidate power in the hands of the state. The language of *morom* or affection became an important lens for foothill residents to describe their experiences of the multiple sovereign entities and the power structures.

CHAPTER FOUR

The *Haats*

P EOPLE advised me to get to the *haats* early. "You have to get there before 6 a.m. to catch the action," explained Jayanta, the *haatkhawa* (tax collector) of the Rajabari weekly market. Activities started early at the *haat*. By 5:30 a.m. on that November morning, Jayanta and his assistant tax collector were collecting taxes and organizing the traders who emerged from the mist to set up their stalls. Trucks loaded with traders and goods, overcrowded buses stacked with people and sacks, hand carts brimming with bananas and vegetables, bicycles balancing baskets of chicken and roast pig, and cultivators with bamboo baskets filled with papayas and ferns perched on their heads arrived at the market. The retreating monsoon rains turned the space into a muddy plaza. Bicyclists and handcarts got stuck in the slush, and laborers lined up to offer their services, as trucks and buses rolled in with clothing and food. Sellers carried sacks of potatoes, garlic, rice, and vegetables on their backs into the market area.

Along with the buyers and sellers, children also came to the market. Some of them came with friends in school uniforms; others followed their parents, grandparents, and relatives. The small children tried to keep up with their parents as they hurried along the stalls. At each puddle, parents effortlessly lifted the children over it. Teenage girls and boys carried bags of vegetables and goods and helped their parents. Beside a cobbler, a group of students in school uniforms stood in line—some to fix their bags, others to mend their shoes.

These weekly markets cater to hundreds of villages far from urban centers and markets. They function as an assembly point where commoditized goods and sentiments shape people's ability to adapt in entangled social worlds. The interplay of exchange and commerce in the weekly markets captured the richly layered world of secrets, taboos, and transgressions in

A wheel of fortune game at the Naginimora *haat*

the foothills. Recalling the origin of the *haats*, village elders said, "Little, very little. We were little." They looked at their hands nostalgically and talked about their experiences. A state administrator in Assam who complained about the unruly behavior of buyers and sellers at the *haat* described them as "a natural happening" and continued, "These are from ancient times, and we cannot just uproot these institutions." But, apart from a few *haats* dating back to the early twentieth century, most were relatively new. Significantly, the *haats* had arisen in the wake of land conflicts. For example, in 2009, after a conflict between a Naga coal mining village and an Assamese village over a boundary issue, the regular *haat* shut down for several months. Buyers and sellers began to meet secretly in a neighboring village to trade. Gradually, the neighboring villagers noticed the activity and arrived there. This was the origin of the Bekajan *haat*, a weekly market that assembles outside the wall of the Oil and Natural Gas Corporation unit in Borholla.

The *haats* are a vital sign of the social life here. The market transactions highlight how power relations are reinforced and amplified through touch, smell, and taste. Thus, the *haats* are not solely about the produce. Visually, they resemble rural markets anywhere in the world. Stephen Gudeman and Alberto Rivera (1990: 25) underline how the basis of the rural economy—the household—in the highlands of rural Colombia is dependent on the relationship between the peasant and the land. This connection is founded on social relations, and not solely on the produce. The social world of the Colombian markets that Gudeman and Rivera present is homogenous, integrated, and clean. By contrast, the *haats* scattered across the foothills function as zones of social interaction mediated through networks of power and alliance founded on extractive activities.

Although products such as mats, tea, and clocks are sold at the *haats*, the market is swamped with the color and smell of food being sold and filled with the noise of the crowd. Items sold at the *haat*, such as bananas, chickens, and pumpkins, embody the social relations here. Sensory feelings such disgust, fear, or excitement are localized and become visible. Therefore, beneath the buying and selling of goods at the *haats*, these markets mask the social signs and meanings of the foothills. As Henri Lefebvre puts it, "Things and products that are measured, that is to say reduced to the common measure of money, do not speak the truth about themselves. On the contrary, it is in their nature as things and products to conceal that truth" (1991: 80).

EXPECTATIONS

"Are you lost?" I asked a young boy in school uniform. He appeared anxious as he sat alone in a tea shop. Madhuja Hazarika was eleven years old and lived in a village near the Rajabari *haat*. He had arrived with his father earlier that day with a basket of bananas. His father left him at the tea shop and entered the market to sell his produce. "My father will come back and take me to a cloth stall," Madhuja said, frequently scanning the crowd for his father. When I left the tea shop later that afternoon, Madhuja was still waiting for his father to return. The market was filled with expectations: finding a new shirt, a good bargain, high profits, delicious food, or reunions with families, neighbors, and friends. It was a place where one could fix old shoes and flat tires, or mend flashlights and broken umbrellas. It was also a place where people came to experience different tastes and indulge in the excitement the market generated.

Attracting hundreds of villagers who had no access to big towns and cities, these *haats* brought their sensual pleasures and thrills to the villages and towns. Shuffling feet, ringing laughter, passionate haggling, aromatic rice beers, spicy pork dishes, crispy fried fish, clucking chickens, and mooing cattle all came together to produce the sounds and smells of the foothill weekly market. Different ethnic groups in their *mekhalas* and *chadors* exploding with radiant colors clung to their prejudices about one another or momentarily left them behind as they participated in the social frolicking at the *haat*. Ahom women in their silk attire crowded the stalls selling jewelry and cosmetics. Adivasi women in colorful saris with brilliant ribbons in their hair browsed through sandals and shoes. Young boys from Assamese villages flaunted their trendy jeans and ate street food. Cultivators from Old Tssori and New Tssori villages in Nagaland carrying beautiful bamboo baskets over their shoulders leisurely strolled around the stalls. Village priests, coal traders, security guards, schoolteachers, farmers, council members, and policemen all assembled at the *haat*. When they saw a familiar face, they exclaimed, "Oh, you are also here!" and rushed toward each other with outstretched arms.

Goods at the Rajabari *haat* were carefully separated: fresh vegetables, clothes, poultry and meat, utensils, sandals, fermented fish, and clothes. At the entrance of the *haat* sat the cobblers, cycle mechanics, blacksmiths, and medicine men. In the next lane, colorful nail polish bottles, lipsticks, and jewelry were laid out for inspection. A group of young girls, hypnotized by brilliant hair ribbons and bells in a stall, ignored the stray dogs that sniffed their feet and licked their hands. It was the clothes stalls that attracted the largest numbers of buyers. "There are no clothing stores in the Rajabari area," one resident said as she picked out woolen cardigans for her household. Not every item came with fixed price. Produce from the Naga villages were weighed, checked, and given an exchange value. As goods were translated into monetary values, tensions ran high.

LOSS

"It is a loss for you," Nrilo reprimanded her grandfather. He had just sold four pumpkins for ten rupees (eighteen cents) at the Nagabat *haat*. When an Assamese customer said to him, "You cannot carry it back, can you?" the old man quickly sold his produce. Buyers from Assam constantly reminded the Naga traders about the risks of carrying the produce back to the hills. These uphill journeys took three to five hours depending on the season. Explaining the logic of profit and loss, Toshi, a Naga cultivator selling his produce, said, "The present trading practices between the hill and plains are this: for example, if there is a good chili harvest and we sell for ten rupees per kilogram, then the traders from the plains get together and say, 'This is a good harvest, do not buy the chili for ten rupees per kilogram, pay only four or five rupees. Better still, do not buy their produce, so that they will lower their price and we can buy it cheaper.'"

Such frustrations were expressions of what it meant to traverse the foothills. These *haats* featured not only fun and excitement but bitter accounts of loss and dismay. Amenla, a Naga trader from the coal mining village of Anaki, sold her produce at the Singibil *haat* for many years. In 2007, she stopped going. She was tired of the journey and said, "The biggest loss for us is that all the traders in the hills carry their produce on their heads." The absence of roads caused Naga traders to incur losses, and traders from Assam took advantage of this to make handsome profits in the upland Naga villages. For example, traders in Assam factored labor and time into the price of the produce every time they climbed the hills to sell their goods. Lima, a frustrated Naga trader, said, "The plains traders always profit, even when they are selling the same chilies back to us. This is because they calculate their carrying charges, the time they spent carrying it, the service tax, among other things." This was not the case when Naga traders from the upland villages came down to the *haats*. He explained how such negotiations unfolded: "They say our hill chilies are not spicy and so they will pay half of the current market price. Next time when we go down to the market with the chilies, they say, 'Oh, last time we paid you too much, we paid you ten rupees. So, this time we are willing to buy your produce only for eight rupees per kilogram.' Once we take our produce to the foothill markets, we have to sell it there. How can we bring it back?"

Abemo, a resident from Bhandari town, described the transactions at the *haats* as "senseless." He said that traders from Assam hoarded the produce and sold it back to them at higher prices. Describing the irony, he commented, "We are happy to buy it back, we are very happy to buy it back!" These profit-making schemes, according to Abemo, mirrored tactics of the state of Assam: "They think how they can harm us by running us into huge losses. Just like the traders from Assam cheat us, the state of Assam will, given a chance, take away everything we have in Nagaland for free."

Yet, the separation between "us" and "them" was porous. Coal traders from the Naga and Assamese villages worked together. Similarly, cattle traders and brokers from Nagaland and Assam worked together. For example, cattle traders from Assam and cattle brokers from Nagaland who delivered huge consignments during festivals and weddings across the foothills, were known for being unscrupulous. Chumbemo described the Naga cattle brokers from his neighborhood in Bhandari town: "Whenever we go down to buy a cow in the foothill *haats*, especially in Merapani, our own people will go down and become brokers between us and the sellers in Assam. They will

Naga traders at the
Nagabat *haat* (Assam)

hike up the price of the cows and share the profit with the plains traders."
This story about cattle brokers disrupted the image of the gullible Naga
tribesman from the hills versus the cunning plains trader from the Brahmaputra Valley, and revealed a social life of exchanges and negotiations based
on alliances and partnerships.

During the harvest season in November 2009, women traders from Nagaland arrived at the Nagabat *haat* with sacks of chilies. One day, I found myself
at the center of a chili deal. "I am not in charge of payments!" a young trader
explained to the Naga women. There were chilies everywhere—radiant green,
white, red, and black chilies heaped on the ground, nestled in bamboo baskets, packed in jute sacks, wrapped in shawls and scarves, and stashed in
shoulder bags. The air was thick with the aroma of fresh chilies.

"We sell it for fifty rupees per kilogram, but after it reaches Jorhat or
Golaghat, it will become one hundred to two hundred rupees per kilogram,"
the women traders told me as they waited for their payment. When the
produce agent from the Assamese town of Golaghat failed to include the
produce of one of the traders, a squabble broke out between the agent and

the women. "You should have taken a piece of paper to write down the quantities and the amount," the women argued. The trader whose produce remained unaccounted for claimed she had given the agent sixteen kilograms of chilies, but he dismissed her claims. As the argument became heated, the women made an additional demand. They insisted that the produce agent not only should pay the woman he had left out but should make sure that all of them were paid the same rate. The agent responded, "That is not the point. I have fixed different prices with different traders. That is the way I operate." The price of chilies at the *haat* depended upon the negotiating skill of each trader. For the same produce, some got forty rupees a kilogram, others forty-five, and the most persuasive traders got fifty rupees. The price of the goods was dependent on the bargaining skill of each trader.

The women traders asserted that everyone should be paid fifty rupees. Surprised at the behavior of the women, the produce agent and his assistants raised their hands to signal that they refused to negotiate. The women stood firm. Unless they were all paid the same rate, they would take back their chilies. Eventually, the agent paid a standard rate to all the traders, including the woman whose produce was previously unaccounted for.

CONFLICT

Sometimes squabbles inside the *haat* took on antagonistic overtones. Given the nature of land conflicts in the foothills, a personal fight between Nagas and non-Nagas inside the *haat* often turned into a border fight between Assam and Nagaland. The consequences were immediately felt in the *haat*. In 2005, Naga traders were banned from entering the Singibil *haat* in Assam after a trader named Atul Rajkhowa from an Assamese village went missing. People from Rajkhowa's village suspected that a Naga village was responsible for his disappearance. Soon after, the Assam police erected an additional checkpoint to stop Naga traders from entering the *haat*.

When I went to the Singibil *haat* in 2009, the adjoining villagers said, "Meeting patibo lage" (There should be a meeting), for resuming trade with the Naga villages. One day, a member of the Singibil village council invited me to Yonlok, one of the Naga villages that had been banned from the *haat*. On our way to the village, we crossed several trucks loaded with coal and timber heading toward Singibil village. The heavy traffic suggested that extractive trade between Naga villages and Assamese towns continued despite the sanction on Naga traders.

People at the Singibil *haat* said that Atul Rajkhowa had been on his way to a Naga village in the upper elevations to buy betel leaves and collect a debt.

A few days later, news of his death reached his village. Some traders said that people from Nagaland had sent a message that Rajkhowa's body was in a Naga village, but they were unable to go because there was a curfew in the foothill border due to a land conflict. As the conversation went on, some insinuated that the deceased should not have gone to the hills if he was a law-abiding citizen. They raised doubts about his connections in the Naga villages. At that point, Rajkhowa's neighbor said he was a good man. "He was not a troublemaker or a thief. He did not even drink," the neighbor commented. Rajkhowa was a man with no vices. His only fault was that he went to the Naga villages in search of betel leaves, a prized commodity that fetched a handsome profit in the *haats*.

A group of young traders from Singibil spoke about seeking revenge, but in the same breath they reminisced about how Nagas were also friends. Singibil village, like the neighboring town of Gelakey, was connected to the coal mining trade here. Many youth from Singibil were coal traders and made regular trips to the Naga coal mines in the upper elevations of the foothills. Therefore, memories of friendships seemed to increase the feelings of betrayal. Several residents suspected that Rajkhowa was killed because of the border dispute between Assam and Nagaland. Rajkhowa's family and friends connected his disappearance in 2005 to an incident that had taken place a year earlier.

Apparently, social relations in the Singibil *haat* became tense in 2004 after an Assam Police Battalion checkpoint was set up outside the weekly market. Soon, state officials from Nagaland came down and warned the Assam administration not to encroach upon land that belonged to Nagaland. A village elder reflected on the behavior of an officer-in-charge (OC) from the Assam Police who was stationed at the new checkpoint. The OC began to harass the Naga villages as well as the traders who came down to the market.

The police checkpoint was put up to resolve a land dispute between the Naijung tea plantation, which was under the jurisdiction of Assam, and the village of Simising in Nagaland. The residents of Singibil village described how the events unfolded. When the tea plantation started expanding toward the uplands, the Naga village asked the management to pay a tax before extending the plantation onto village land. The plantation manager reported this to the Assam police as a case of extortion, and the administration in Assam arrested and imprisoned the village chairman. "In those days the situation on the border was tense, so the Assam government set up an Assam Battalion camp near the Singibil *haat*," the resident continued. "The Assam Battalion closed the path Nagas took to come to the Singibil *haat*."

This was not the first time the market in Singibil was affected by the border conflict. The Singibil *haat* originally took place in the upper elevations of the foothills, but it was moved to the village after a conflict in the 1960s. "There, over there. Do you see those hillocks? The *haat* bazaar used to sit there before it was moved near Nogan Gogoi's house in Singibil village," Rahman said. It seems Nagas came down to attack Assamese villages during the land conflict, and villagers came running to Rahman's house shouting, "The Nagas have come from the hills again!" This statement seemed to reiterate the stereotypical image of Naga tribes as a warlike savages who regularly raided Assamese villages. During this period, the Assam government posted the Armed Police and the Special Reserve Battalion from Assam near Singibil village.

The *haats* were filled with stories. Accounts of boundary disputes were an integral part of the markets. When I inquired how certain ethnic groups had been banned from the *haat*, villagers in Assam could not name any legitimate authority that had passed these regulations. Referring to the 2005 incident that banned Naga traders from the Singibil *haat*, the tax collector said, "Someone in the *haat* must have told them not to come because there is trouble. Something might happen anytime." The conversation about the ban sparked a heated exchange between an Assamese trader and the tax collector:

> TAX COLLECTOR: You asked Nagas not to come to the *haat*.
> TRADER: Who said so?
> TAX COLLECTOR: You did.
> TRADER: If someone said that to the Nagas, I do not know. But I did not tell them that. Whoever has told the Nagas not to come said that one of their people has been killed in their hills. There should be a meeting to settle the matter.
> TAX COLLECTOR: But what can we do now?
> TRADER: Then who will come and show the guilty person?
> TAX COLLECTOR: The Nagas will point out the murderer.
> TRADER: Nagas will never be allowed to step foot in this *haat* again. No, they are not coming here. We are also human. If we see a Naga, we want to kill them or beat them up.
> TAX COLLECTOR: Hey, you are talking very big!

Eventually the argument subsided. The tax collector announced, "The Nagas should come down to the *haat* again. The economy will be better if they start coming to the *haat*." The tax collector was concerned about the loss of revenue since the Nagas had been banned because there was a steep decline in revenue. Twelve Naga villages who came down to trade at the Singibil *haat* had stopped coming down after 2005. In December 2010, administrators from Assam and Nagaland held a meeting in Singibil village. Traders from Naga and Assamese villages, including Atul Rajkhowa's family, attended the gathering and welcomed the Nagas to the *haat*. Yet not all hostilities could be reconciled because these markets were also spaces where people defended their cultural practices and beliefs as superior to others. Experiences of humiliation and transgression were recurring themes in conversations about the *haats* in the foothills.

HYGIENE

A rich Assamese coal trader at the *haat* scoffed at the absence of medical facilities and schools in the Naga villages. Exclaiming, "Chi!"—an expression that denoted disgust and resentment—he continued, "I do not know how they live like this," and explained that conflicts in the foothills were mainly due to the different hygienic practices of the hill people and the plains people. This matter, according to him, went beyond the security forces and the state administration. He defended the presence of the security forces in the foothills: "They [security forces] are here to protect the public. They are here as a show of strength. However, if you and I cannot get along, the police and the military are not equipped to bring us together. They cannot resolve our conflict. They can only maintain the status quo. Everyone is equal. The Assamese are equal, and the Nagas are equal. But if one is hygienic and clean, then it is not a problem to get along with people."

The reference to hygiene reflected the affluence of Assamese households in contrast to the poor Naga traders and asserted a superior civilizational position for those who lived in the Brahmaputra Valley. It seemed that cleanliness and hygiene were markers of refinement and progress that elevated valley residents above the Nagas.

An old Naga trader described how buyers from Assam were careful not to touch him: "They did not even allow us to enter their houses. Even in the market, they dropped the money in our palms and made sure they did not touch our hands." The experiences of the Naga trader show how economic practices were founded on social values and morality. The emphasis on proper hygienic practices as the basis for the Nagas and the Assamese to "get along" demonstrates how disgust significantly frames the social order here. Perceptions about hygiene emerged from the trade and exchanges at the *haats*.

Participants recalled the weekly foothill markets through taste, sound, smell, sight, and touch. Fresh produce such as yams and pumpkins not only invoked the aroma and taste of food but generated insightful conversations about social relations. While sellers from Assam gathered information about various ethnic groups and their preferred foods, Naga traders described their awareness of the Assamese preference for red chilies. One trader told me, "They [households from Assamese villages] love the taste and are drawn toward it. They make pickles, and ripe red chilies are the best."

The various tastes of produce from Nagaland and Assam were emphasized in the *haats* and became objects of desire. When people spoke about their longing for certain kinds of herbs, fruits, or vegetables that were found only in the *haats*, the boundary between the hills and the plains ceased to exist. The flow of goods between the upland Naga villages and the Brahmaputra Valley was continuous. Many Assamese households consumed large quantities of areca nuts with lime and *paan* (betel leaf). Although areca nuts and betel leaves grew abundantly in the plains of Assam, the Assamese villagers preferred the produce that arrived from upland Naga villages.

For instance, Mr. Saikia from Sibsagar district told me that he could not eat the *paan* leaf that grew in the plains of Assam. He opened his mouth wide and displayed his tongue flecked with areca nuts. "It has to do with my tongue," he explained. "My taste buds demand produce from the hills." Similarly, Mr. Kithan said, "Everything in the mountains tastes so good." He came from a village called Champang ONGC in Nagaland, where the ONGC had explored for oil and gas until 1994. Although he seemed to emphasize that food from his farm tasted better and was of superior quality, in reality oil spillage and seepage in his village had damaged land fertility and caused health hazards such as renal problems. In addition, due to the oil seepage, *jhum* cultivation was affected because farmers were afraid to clear the land by burning due to the oil spills.[1] Yet, Mr. Kithan continued, "In the hills, the vegetables are tasty, but in the plains vegetables are bitter and tasteless. Take oranges—they taste different when grown in the plains and hills!" His description of vegetables from the plains as "bitter" was more about favoring his ethnic identity as a Naga over that of the Assamese and other social groups from Assam.

The *haats* do not represent the consumption culture of the foothills. Delicacies such as red ant nests and woodworms that many ethnic groups relish are not sold in these markets. During the monsoon, frogs from the paddy fields became part of everyday cuisine, and in winter the streams and rivers supplied an array of insects, crabs, and snails for sumptuous feasts. Unlike foods such as pork and beef, which occupy an unambiguous slot in the world of dietary taboos, the transitional ecology of the foothills offers a diversity of livestock and aquatic life that disrupts the dominant taboo categories. A Naga resident explained to me that Naga traders only sell what "they" (villagers in Assam) will eat and not what "we like to eat" (referring to the Naga villages). For this reason, woodworms, red ant nests, beehives, mushrooms, locusts, and snails were missing from the *haats*. Naga traders at the *haats* focused on ginger, pumpkins, yams, and edible ferns to attract buyers from Assam. Dominant food taboos were observed in these *haats*. For example, upper-caste groups like the Brahmins adhered to strict taboos against eating meat and bought solely vegetables at the *haats*.

The foothills are part of a larger story about migration, conversion, and settlements. Girin Phukon, a local historian from Moran, described how the Tai-Ahom rulers in parts of Eastern Assam (commonly referred to as Upper Assam) converted to Hinduism during the seventeenth century but did not adhere to dietary restrictions. They continued to eat beef, pork, and buffalo at social functions such as funerals, festivals, and weddings. Describing the history of food habits in Ahom villages, Phukon notes that Ahoms (who had converted to Hinduism but were not Hinduized) consumed pork and fowl, and drank rice beer, unlike their Assamese Hindu neighbors.

Phukon (2010: 286–87) further notes that "duck, pork, young bamboo shoots—ground and fermented, dried fish *hukati* (*pai* niu), . . . crabs, maggot, woodworm, frog, and many varieties of insects both of land and water" served as regular cuisine in Ahom households. Girin Phukon's (2003) work on Ahom dietary history highlights that imposition of dietary restrictions is connected to power relations and the assertion of identity along the foothills. Yet, the *haats* continue to capture the social world of the place.

One day, I quickly wrote down three simultaneous conversations between buyers and sellers in adjoining stalls at the Nagabat *haat*:

> Stall 1: Bengali and Ahom poultry vendors talked to Naga buyers: "Are you buying the duck to feed the church pastors for a good harvest?" The vendors appeared excited while asking their Naga customers about village church activities during the harvest season.

> Stall 2: "Are you going to have a feast today?" Mishing butchers teased a group of Ahom women who were buying pork from them. Next to the Ahom women, a Naga couple arrived at the

stall and waited their turn to check the meat. The young Mishing butchers smiled and began to solicit the Naga couple: "Hey, when is your Naga Bihu [harvest festival]? Come to us, all right?"

Stall 3: "You are going to make some tasty curry this evening. Have you got your weekly payments? Do you want to buy in bulk?" Bengali dry fish vendors sitting behind mountains of dried salted fishes, prawns, and dried chilies chatted with Adivasi girls buying an assortment of the aromatic goods displayed at the stall.

These exchanges indicate how knowledgeable people in the *haats* are about each other. Friendships often grew out of these intimate engagements. At the Nagabat *haat*, Naga women traders frequented a regular spot to drink tea and take a break. The tea stall belonged to Munna, a thirty-year-old tea vendor. Covering six *haats* every week, Munna's mobile tea shop consisted of a small kerosene stove strapped on his bicycle and a wooden tray with an assortment of buns, cakes, and breads in glass jars. The Naga women traders introduced me to Munna: "He knows how much sugar we take, he is our friend," affectionately teasing him. Munna's mobile tea stall was a secure place where the women could take a break and store their goods. On *haat* days, after they had sold their produce, they brought bags of rice, duck, and bamboo mats and asked Munna to keep an eye on their belongings. They insisted, "Itu bhi rakhibi deh" (Keep this as well), and hurried away to continue their shopping. Women returned at regular intervals carrying more items in their bamboo baskets—oil, kerosene, medicine, and blankets—and stacked them at the tea stall.

This familiarity also extended to sellers inside the *haat*, like Romizuddin Ahmed, who sold medicine from his mobile pharmacy. He came from Golaghat town and had been visiting the Nagabat *haat* for two decades. The Naga women traders fondly addressed him as Dr. Romeo and bought vitamins, ointments, and deworming tablets from his stall. When I visited Romizuddin's house in Golaghat town during an Eid celebration in 2010, I noticed that he owned a pharmacy and a large concrete house. As I joined his family to celebrate Eid with beef curries and rice, I inquired why someone like him, comfortably established, would regularly visit the Nagabat *haat*. He could surely enlist an assistant or employ someone to handle the mobile business. Romizuddin told me, "I am sixty-five years old, and I do not need to do this anymore, but I like to meet my friends. I like to set up my medicine stall and talk to the people." People were drawn to these markets to meet friends or

Romizuddin Ahmed in his medicine stall (Assam)

make new friends. As much as hostilities and violence were real in the *haats*, so were relations of affection and long-lasting friendships.

A SMELLY CLEAN FRIEND

Molong Ali described how he found his best friend at a *haat*: "Nagas are good, but it is the plains people like us who spoiled them. We taught them all bad things." Ali's clearest memory of his best friend Wangcho, whom he had met at the weekly market, was the smell of his body. "Sometimes if a Naga sits down in a particular area, people are not able to sit around that place due to the smell. Nagas do not shower often. But let me tell you whatever they say from their mouth is all clean—their minds are clean," he said. Ali declared that their food was delicious although the recipes were simple, drawing an analogy between the people and the food they ate. Both were simple, yet hospitable and endearing.

Ali's distinction between Wangcho's clean mind and the bad things plains people from Assam taught the Nagas in the hills might be read as an attempt to infantilize their friendship. But it was interesting how Ali detached the notion of cleanliness from corporal practices of cleaning the body so that body odor and disease could be prevented. Instead, he selected a different framework of cleanliness to describe the qualities of his friendship with Wangcho and extended those traits to the Naga people. To reinforce his point about the "minds" of the Nagas being clean, Ali said, "The food they make is so good. Their houses are designed in such a manner that the first room you enter in the house is the kitchen. I still tell my daughter-in-law that she will never be able to make yam curry with mustard leaves like the Nagas prepare it." He reminisced about a Naga household that had prepared elaborate meals for him during his many visits there:

> They cooked the mustard leaves. First, they thoroughly cleaned the leaves with water and then scrubbed the pots before they started cooking. I wondered. They do not clean their bodies so often, but that was not the case with the food they ate. Then they filled the pan with water and added the mustard leaves. They cut up a few green chilies and threw them into the pot, added salt, and boiled the leaves. Next, they cleaned the fish properly, chopped it up into small pieces, and put it in a pan. They added water, salt, green chilies, and dry bamboo shoot. Finally, they cooked beef for me in the same style because I did not eat pork. The food was so tasty! They boiled everything, and it tasted so good. The more they boiled the food, the better it tasted.

As a coal and timber trader, Ali spent long months in the upland villages. During the summer, he camped in the forest, but he lived with his friends in the Naga village during the winter months. Ali spent long hours near the fireplace, so his reflections on the cleanliness and hygienic habits of the Naga people centered on food and its method of preparation. Although his friend and the villagers were "smelly," he described how they thoroughly washed the food and cooked it well.

He laughed out loud when recollecting a particular story. Ali recounted that one day he went to meet his friend Wangcho. When he arrived in the hilltop village, he learned that his friend was not home. After the household served him dinner, Ali went to bed. "In the middle of the night, I woke up because I could not breathe. I was suffocating." Ali looked around the room, but it was dark. Gradually, his eyes adjusted to the darkness, and he understood what had happened. He found Wangcho's shawl on top of his blanket. He realized that his friend had put his shawl over the blanket because it was a cold night. Molong Ali could not tolerate the smell and threw off the shawl. The noise woke his friend, who was sleeping in the other bed. He looked at Molong Ali and commented, "Oh, so you do not like the Naga smell! I thought you would be cold—that is why I gave you my shawl!" They both burst out laughing. "When Wangcho died, I cried a lot over that old friend of mine," Ali said.

CONCLUSION

The *haats* are important sites that reflect the social and economic life of the foothills. Connected with myths of origin and local ties to place, people regard them as important signposts of historical, political, and cultural association across the foothills. Yet these *haats* are located in precarious sites: inside a disputed tea plantation between Assam and Nagaland, on the fringes of a coal mining village, outside the walls of an oil and gas gathering station. Therefore, during land conflicts they shut down, or specific ethnic groups are banned from entering the markets to trade. Power relations and hierarches became visible through the economic transactions in the *haats*. Across the foothills, traders from Assam dominate the economic transactions in the *haats*, but residents also derive meanings and attach values to the produce, invoking connections. Taboos and transgressions as well as the taste of food such as chilies, areca nuts, and vegetables in the *haats* are categorized as belonging either to the hills or the plains. But this knowledge is produced in the foothills.

CHAPTER FIVE

Extractive Relations

"ARE we there yet? Are you there yet?" Pradip was restless. As we traveled through the Gelakey tea plantation toward a Naga coal mining village, he became increasingly agitated when he failed to see any clear demarcation between Assam and Nagaland, asking, "So where is the border? Where it is it? Whose lands are those? Is it Assam's land? Are we still inside Assam?" The coal traders taught him how to read the foothill landscape. Bhai, one of the coal traders, said, "As long as we are traveling through the tea plantations, it is Assam land. Once we enter the higher elevations of the foothills and come to the *jhum* fields, it belongs to the Naga people from the hills." The *jhum* fields and tea plantations merged at several places, creating more confusion for Pradip. He said, "How can it be? How can this land belong to Nagaland?"

Disputes about boundaries took place at many levels, but those who lived in the foothill border negotiated both social and physical borders every day. Although conflicts regularly shut down transactions at the *haats*, people traveled and entered into various trading contracts and alliances. They did not wait for periods of stability and peace to trade. In the coal mining villages across Nagaland, these social relations were founded on the networks of coal traders, reciprocal labor relations, and adoption practices. What would it mean to see the foothills through these everyday practices? A Naga elder commented, "There was so much talk about friendship and goodwill, they left out the boundary talk during the discussion." Resource conflicts among ethnic communities escalated during particular seasons, such as harvest time, when the grains had to cross several security checkpoints, or during the planting and weeding season, when tools and implements including fertilizers had to travel to the farms.

These anxieties were heightened in the coal mining villages because mining was a seasonal activity there. Unlike Oil and Natural Gas Corporation activities and tea plantation operations, which continued all year round, the coal mines were closed during the monsoon and required substantial logistical coordination during the dry winter months (November–April). Given the pressure from the government of Nagaland to ban coal mining and the resistance from the Naga villages, the coal mining activities centered on establishing special relationships and ties. Coal from the Naga villages, like Assam oil and Assam tea, was given an ethnic title (Naga coal) and the extraction process involved managing kinship ties and social relations in the villages. Despite hostility between the government of Nagaland and the coal mining villages, mining generated new forms of relations between Naga and Assamese villages in the foothills.

SEASONAL FRIENDS

As our jeep crossed *jhum* fields and coal depots, the coal traders who worked in the Naga villages talked about managing friendships. "It is tricky, this friendship thing here. Only when the coal season begins is there a real sense of closeness. We bond during this time and we are friends. We joke, drink, laugh, and we trust one another," they said. When I asked the traders why this was the case, one of them said, "You see, it is like this because there is frequent movement of people and cars during the coal season. We are always bringing salt, rice, and vegetables and doing favors for them." A second coal trader elaborated: "After the coal season, it is difficult for us to frequently visit the hill people and their homes. We have to do our own thing, and that is when there is a communication gap all over again. We hold back our thoughts and become apprehensive about how to mingle with them. We become cautious and afraid of one another."

Such conversations about seasons and friendships were routine. These social relationships were so embedded in resource extraction that they thrived during the long dry spells from November to April. During these months, coal was the most visible item in the foothills, leaving a trail of dust and soot across the jungles and villages. With the beginning of coal season, traders competed to win over networks of landowners in the Naga villages. For example, coal traders from Assam often kept track of each other and gathered information on who came to the Naga villages for business and trading opportunities. When I traveled with coal traders from Assam to the Naga villages, they frequently turned to scan the license plates of the vehicles and note the people inside. When they recognized the traders, they commented, "What are they doing here? They are crooks." These comments captured their anxieties about the friendships established during the coal mining season.

Similarly, Naga villages kept track of who came to the village with business deals to dig for coal. Only Naga landowners could engage in mining, so traders from neighboring villages, whether they were Assamese or Naga, had to enter into joint ventures with the landowners who controlled the mines. Consequently, friendships were entangled with resources, patriarchal property regimes, and administration that developed differently in the hills and the plains. Assamese and Ahom traders often repeated, "Everything below the ground in Nagaland belongs to the people, but in Assam it belongs to the government." They were referring to Article 371A, a special provision under the Indian Constitution that granted the Naga people rights over land ownership and natural resources in Nagaland. For traders from Assam, "the people" meant Naga men who owned resources. However, there were overlapping claims about who owned the land and who was the legitimate authority in Naga villages.

The parties negotiating business deals in coal, timber, or rubber ranged from individuals to the biggest companies in Nagaland and Assam. These collaborations took place with different parties. In the Naga coal mining villages, coal exploration started with the consent of the landowners. However, permission to start mining activities came from the Naga insurgents, after the trading parties agreed to pay the "Naga national tax" to the Naga armed groups. This tax amounted to between 15 and 25 percent of profits. To get the best deal, it was considered advantageous to include a Naga insurgent's household member on the trading team. Involving insurgent family members facilitated safety and security for the operation. Naga insurgents were recognized as powerful figures who could make possible the extraction of coal.

Once the coal trucks were loaded at the mining site, the vehicles required permission from the Nagaland police to leave the village. This often meant making payments at various checkpoints, where Naga insurgent groups often negotiated with state officials on behalf of the villages. As the truck left the village, the owners had to pay a state tax to the Directorate of Geology and Mining, including a tax levied by the village council. Finally, negotiations with the Assam police began at the border checkpoint. Here the coal traders from Assam used their social connections and networks. Under such circumstances, friendship became a complex relationship that was necessary to navigate the movement of resources like coal in a precarious landscape with multiple competing power networks.

Tali, an octogenarian Naga trader, explained that friendships and solidarities in the foothills were founded on fear and anxiety. He recollected how people in Assam reacted when they saw Naga traders: "People in Assam ran away. They feared us. We stopped behind their paddy fields and cooked, ate, and took rest. They could not tell us anything. Not because they loved us, but because they feared us. It was not love. Nagas are cannibals—that's what they thought!"

Tali's awareness emerged not from official documents about the borders but from daily practices, expressions, and classifications. When villages appeared in the foothills and people started trading, they had to interact with one another. Activities such as grazing, foraging, and fishing meant that access to forests and meadows overlapped. In addition, several households had adjoining farms and shared common entrances and approach roads to their villages. Foothill residents embraced the language of friendship and familial ties to sustain their social ties and networks in this place.

Every household in the foothills was involved in trade, but that trade was not based on formal agreements. When I visited households of coal traders in Gelakey, people arrived in groups of twos and threes from the neighboring Naga villages. They drank tea and engaged in mundane conversation. These moments appeared as what Marcel Mauss called "fleeting acts of hospitality" (1990: 81). For Mauss, societies adopt these acts of hospitality or practices to overcome reservations and enter into reciprocal relations with their fellows. These alliances are not made by choice. Naga landowners and coal traders from Assam had to transform fear into mutual trust and friendly alliances to survive here. In the coal mining villages and towns of the foothills, during these moments of hospitality, people shared information about families or ailing neighbors and friends. Mundane as they were, these conversations constituted an important part of establishing trust in making alliances. It was often by word of mouth over tea and food that contacts and new coal sites were discovered.

FRIENDS AS WITNESSES

Once contacts were established and coal leases drawn up between traders and Naga landowners, the performances of friendship were impossible to ignore. The one-page English language agreement was signed, and the trading partners—the landowner and the trader—urged each other to read the document. Many were in English and contained similar lines and clauses, suggesting that residents had copied previous agreements. They looked like legal documents because all had itemized clauses, but the clauses had been greatly altered as neighbors and friends reworked the original wording.

Anthropologist Annelise Riles (2006: 4–5) states that documents can create an "ethnographic response" and emerge as "cultural texts." "Part art and part technique, part invention and part convention," the documents are part of the ethnographer's work. In Naga coal mining villages, the performances around the documents served as an important practice of trade and alliance. They also provided insight into the parties' social worlds. Spelling errors, illegible handwriting, and added phrases made each agreement a distinct cultural text.

The documentary performances invoked various authorities and institutions, including village councils, the landowners, and even the police officials to be paid for moving the coal from the mines. The role of the parties signing the agreement lay, among other things, in evoking an official aura as the documents were signed and read. In the coal mining villages, English-language documents were required for all kinds of trades and contracts. Yet these documents also produced anxieties and uncertainties. This was less about the life of the document and more about the social lives of the people whose names appeared in agreements. The awkward atmosphere at the signing ceremony captured how agreements determined the course of the relationships that participants had initially forged as friends and family members.

Although the foothills are ethnically and linguistically diverse, the coal agreements between Naga landowners and Assamese traders were all written in English, even though the residents rarely spoke English. English was the official language of the British administrators in this frontier region and is recognized today as the official language of Nagaland. Thus, village councils and customary courts use English official documents even when dealing with neighboring villages in Assam regarding repair work, meeting proceedings, or trading contracts. These village councils and customary courts followed the style of the regular courts and government institutions. As Riles (2006: 12) notes, documentary practices shape the behaviors of organizations, and these English documents created their own mechanisms of operation. The official pads, organizational seals, color, logo, and allotted spaces for names and signatures revealed the process of making the coal mining operation official.

The traders and landowners who signed the documents generally refused to speak English—except when testing their language skills after jubilant drinking sessions. But their manner showed the signing to be an important event. While the content of the agreement was important, it was primarily the rituals and performances that occupied center stage as signing parties strived to affirm their friendship and social bond. The parties induced each other to read the agreement as they signed it, all the while commenting, "Why should I read it? I trust you. You are my brother. You are trying to insult our friendship." Such verbal protestations of trust were often enacted in a jocular manner. Nevertheless, the participants appeared embarrassed when they were persuaded to read a contract, as though reading the agreement compromised their trust and friendship. Underlying this jocularity and refusal to read the contract was a ritual that embodied and affirmed trust between two parties.

The consequences of breaching a written agreement were severe. This interplay of exaggerated sociality and disinterestedness captured how coal traders in the foothills wove what Arjun Appadurai (1986: 10–11) calls "cultural designs" into trading agreements and contracts that appeared to "underplay the calculative, impersonal, and self-aggrandizing features of non-capitalist societies." The Naga landowners and Assamese traders who signed the legal document were wary of each other. Traders could be excluded from trading networks and sometimes physically attacked if the landowner felt that an agreement had been infringed. And the landowners' access to business networks in Assam was likely to disappear if they upset the traders. "There is no middle way: one trusts completely, or one mistrusts completely," Marcel Mauss noted (1990: 81), as he described how human societies that feared and disliked each other learned to trade. Although Mauss was specifically writing about societies where there were no Western legal and economic systems in place, his work is relevant here in the context of exchanges and contracts. Reciprocity in the form of gifts and performances of jocularity constitute the social basis for coal mining. The appearances of stable and secure trading relationships in the foothills mean recognizing these performances as acts directed toward establishing trust and mutual interests in a violent landscape.

SUFFERING THROUGH RELATIONSHIPS

Regardless of the nature of business, trading networks included the entire household. One had to be a *sathi* (friend), *ghor manu* (family member), or *ghor manu nisena*. Foothill residents thought of trading relationships as an extension of opening their homes and providing protection to their trading partners and workers. Thus, the notion of stabilizing a trading relationship was not based on reciprocity alone. These bonds became the basis for inventing family ties and connections through adoption. As economic activities developed in the coal mining villages, Naga households adopted traders from Assam as friends, brothers, or sons. Discussions about trade and business often overlapped with conversations about hospitality and family.

Not all traders found it easy to understand these practices. When I met Sunil, a businessman from Assam, he said that he had begun to trade in coal but found it difficult. It was annoying that his business partners from the Naga villages constantly asked for favors that were "outside the agreement." "Coal trade is like gambling," Sunil said. He was ready to take the risks but described the demands of the Naga landowners as "frustrating." It was unlike anything he had experienced. He described Naga landowners from the coal mining villages: "In the first meeting the landowner will be very nice. He will even wipe the seat for you. After the payment is made for exploring the coal, things are all right until we find coal. Once coal is discovered, the demands slowly start growing. From a bike, it becomes a jeep and so on. When people in the landowner's family fall ill, they ask for help from the business people in the plains to arrange for doctors and medicines."

Previously, Sunil had worked as a tea agent, but he shifted to coal when he found a "line" to the hills. "Finding a line" was a common phrase among traders that referred to ways of becoming part of the trading network. He initially made a profit but later suffered losses. Sunil commented that "the traders from the plains find the police easier to manage than the landowners in Nagaland. The landowners are a headache."

The coal economy in the foothills functioned as a channel of support similar to what anthropologist Goran Hyden describes as "a network of support, communications, and interaction among structurally defined groups connected by blood, kin, community, or other affiliations" (1983: 8). Resources like coal held promises that were not measured solely in monetary terms. Obligations that Sunil found annoying, such as scouting contacts for good doctors or hospitals in Assam, were expected as part of entering into these agreements. This was at the heart of mutual interests. When traders from Assam came to Naga villages, the villagers took them on hunting and fishing expeditions. These activities were outside the clauses in the official agreements, but it was these hospitable gestures they considered when the time came to extend leases.

LEARNING TO BE FRIENDS

"With regard to land ownership in Nagaland, no matter how much the government gives orders to the landowners, the land belongs to the people," Luit announced. He was a coal agent who worked through a well-charted network of Naga insurgents, politicians, bureaucrats, and the village council in the foothills. When I asked him how he managed his "connections and lines" in Nagaland, he replied:

> It is important to cultivate friends—all kinds of friends, ranging from government officials to police, insurgents, and village council members—in the hills because they [government officials] are the ones who control the documents. But the hand of friendship also has to be extended to the people because the land belongs to them! Unlike the land regulations in the states of the plains in regard to minerals, in the hills the land belongs to the people. There is community ownership of property, and any kind of activity in the hills depends on good relations, which can mean many things: hanging out together or going hunting together, etc. In Mon district, it is the *angh* [king] who oversees the system of governance. So, there will be a share for the *angh*, then the village council will get a share, and the student union, but the main parties are the insurgent groups. You have to mainly manage the insurgent groups. Before there were only two groups; now there are many groups. Managing these issues is not easy.

Luit's reflections on managing people, tribal institutions, government bodies, and insurgent groups highlight that for the traders there was no distinction between Naga insurgent groups and government officials. Many Assamese villagers in the foothills considered the Naga insurgents to be the representatives of the state in Nagaland, and Luit seemed to agree. People negotiated these multiple sovereign bodies and authorities in their everyday lives. These authorities portrayed the essence of what constituted Naga-ness as fundamentally attached to land and natural resources. In that same spirit, the identity of non-Nagas was partially defined by restrictions on owning property and an ability to negotiate and manage different political actors in the coal mining villages.

But it was not the Assamese coal traders alone who argued that their presence was legitimate. Various authorities in Nagaland, including traditional councils, Naga insurgent groups, coal mining villages, and the state government, interpreted this constitutional provision in their own favor by claiming to represent the Naga people. For example, Naga insurgent groups claimed they were the true representatives of the people and the leaders of the "Naga nation," while the government of Nagaland claimed status as the legitimate "Naga leaders." These overlapping assertions created challenges at times.

In 2010, Nagaland Chief Minister Rio explained how his role as head of state and his ethnic identity as a Naga created problems: "Some people started saying that the chief minister is anti-national and anti-constitutional. These people who opposed me do not understand. In a democracy, being a Naga, if I don't speak for Nagas, what use am I? As a Naga leader, if I do not speak for the Nagas, what kind of leader am I?"[1]

By invoking the right to speak for the Nagas, the chief minister asserted that he was the rightful representative of the people. During his tenure, a series of consultative meetings about oil and coal exploration took place in Nagaland. On July 26, 2010, the Nagaland legislative assembly, under Rio's leadership, adopted a resolution, The Nagaland Ownership and Transfer of Land and Its Resources in Respect of Mineral including Petroleum and Natural Gas.[2] This resolution specifically focused on Article 371A of the Indian Constitution. It stated that in relation to any "ownership and transfer of land and its resources," the state of Nagaland reserved the right to enforce the "appropriate modalities." Reiterating the position of the chief minister, the resolution noted that the land belonged to the Nagas, but the state of Nagaland, "representing the people shall have the jurisdiction in regulations and development over minerals including petroleum and natural gas, etc." This resolution was rejected in the coal mining villages in Nagaland.

The line between legal and illegal activities, given Nagaland's attempt to bring the coal mining areas under its control, often meant that surveillance of the coal mining villages was intense, and the coal traders felt an increasing pressure to manage various officials. Working as a coal trader in the foothills could be dangerous. But the danger increased fondness for the place. Luit said, "I love Nagaland more than Assam." He referred to the upland villages as "relaxing" and "fun." He went fishing and hunting in Nagaland and spent long weekends with his friends in the forests. Referring to his network of friends in the coal mining villages as "ghor manu" (family members), he said, "I speak Nagamese because my best friend is a Naga. . . . We hang out in the Namtola Gate [a foothill border gate]. Around January 26 and August 15, when there are curfews all over these areas, we go up to the mountains in Nagaland and have fun—just hang out. His house in Nagaland is like my own house."

As Luit described good times with his Naga friends, his account also invoked the reality of militarization. Every year, the government of India declared high alerts across Northeast India on important national days such as January 26 (Republic Day) and August 15 (Independence Day). Because insurgents used these national events to carry out bomb blasts on the ONGC oil rigs in the foothills to send a political message to New Delhi, these dates spread fear and uncertainty across the region.[3] However, the hospitality and the privileges of being welcomed into a Naga family described by coal traders like Luit were grounded in property relations and resource extraction practices.

HOWRI: RECIPROCITY

Several landowning families who tilled the land and practiced both *jhum* and settled agriculture established reciprocal alliances with their neighbors. The result was a distinct practice called *howri* (labor exchange). Bina, an Adivasi cultivator I met in the foothills, explained that *howri* is a *niyom* (tradition).[4] Through labor exchange, *howri* establishes social ties among different ethnic groups in the foothills. During the harvest season in 2009, I followed Aka to her small *jhum* farm in the upper elevations of a village called Champang ONGC. Many residents in the lower elevations had given up farming due to the oil seepage and spills left behind by the abandoned ONGC oil wells. But Aka continued to plant rice and a variety of vegetables. The rice and vegetables she grew sustained the household for a few months; for the rest of the year, she managed by trading at the *haats*. As Aka and I arrived in the field, her *satinis aru ghor manu* (friends and family members) were waiting. When I asked Aka how much she paid the workers, she responded, "They are *howri*. They came because they are family and friends— otherwise they would not have come!" Several women who came for *howri* were family friends or had kinship ties to Aka. Maria, an Adivasi friend of Aka's, had organized the *howri* for her. Maria and Aka's fathers were friends, and the children had grown up together. "They are like brothers," Maria said of Aka's brothers. "We are family," she added. I also learned that Maria was married to Aka's cousin.

The *howri* day had a festive spirit, like the festive parties among peasants from the Cauca Valley in Colombia described by Michael Taussig. According to Taussig, "systems of teamwork, festive labor practices, and reciprocal labor exchange" formed a social organization that was "embedded in a non-market mode of using and sharing land" (1980: 71). The Cauca Valley peasants who performed reciprocal festive labor practices did not concentrate on producing surplus or accumulating profits. Similarly, *howri* services in the foothills were not tied to monetary payment, nor did they focus on profits. The owner offered tea and snacks to the workers, who rendered free labor service in return for a grand midday meal. Everyone referred to one another as *satini* (friends), *ghor manu* (household member), and *ghor manu nisena* (just like a household member). On this day, hierarchies disappeared, and no one, even the landowner or the farmer, behaved like a *sahib* (overlord). The practice of *howri* allowed workers to keep any vegetables they found in the *jhum* field during the harvest. Since it was predominantly a rice harvest, this meant that the *howri* workers had fun finding ripe pumpkins or edible roots that were covered by the rice stalks. The *howri* workers laughed and joked, and young children who accompanied their mothers played in the *jhum* fields.

"This is a practice in the foothills. If a person needs help, a group goes and helps them out," Aka explained. Although her friends came as *howri* for the day, they also worked as daily workers and sometimes returned to her farm as wage laborers. Aka was grateful that Maria had organized the *howri* for her. "Otherwise, it is difficult to get help these days. What to do? We have come to a point where everyone has to fill their own stomach, look after their own family," she said.

Special skills were required for different *howri* activities. The skills needed to clear a forest or work in a *jhum* field were different from those needed for working in a vegetable garden or on a rubber plantation. Although there was no *howri* culture in the coal mines, households in the village rendered reciprocal services in the fields. On Aka's farm, Maria and her Adivasi relatives could harvest rice with ease. They worked gracefully, finding the paddy hidden under a blanket of banana trees, vegetable bushes, tall grass, and wild broom shrubs. As the *howri* visitors worked, Aka explained how her Adivasi friends had learned to work in the *jhum* fields:

> People in the plains do not know how to clear the hills. They
> cannot handle it. They cannot sow seeds in the *jhum* fields.
> Except those who have been coming up to these areas to work,
> many of them do not know how to do it. For some people from
> the valley, they cannot even climb hills. Except for those who are
> in need of work and go out to work for daily wages, the women
> in the valley stay indoors. They do not go out. But these women
> you see are experts; they have been working here and climb the
> hills better than us.

In Aka's eyes, it was the frequent travel to the *jhum* fields and prolonged association with Naga *jhum* cultivators there that made the Adivasi women experts. The skill of the "experts" also lay in their ability to speak multiple languages and grasp various techniques of tilling the land and cropping

Aka serving tea during a *howri* (Nagaland)

carried out in wetlands, highlands, and plantations across the foothills. Aka appreciated her Adivasi relations not only for their ability but also for the relationship of trust and familiarity she shared with them. Her references to "valley women" referred to Assamese caste-Hindu women. Her wording conveyed a disconnect from these women, unlike the bonding and closeness she shared with Adivasi women through their common labor practice.

On the day of the harvest, Aka and her Adivasi friends spoke to one another in Assamese, then switched over to Sadri, and eventually transitioned to Nagamese. Unlike the coal traders' and landowners' English document performances, which were awkward, the *howri* activities were filled with laughter and affection. One of the most important parts of the *howri* day was the food. The breakfast, tea breaks, and lunch were elaborate affairs. Near the granary, Aka laid out tea leaves wrapped in old newspapers, powdered milk in plastic packets, different kinds of biscuits, thick bundles of dried tobacco, and areca nuts with lime and *paan* leaves. Aka and her cousin walked around the *jhum* fields with a big kettle of tea, several white enamel cups strung together on a rope, and a plastic bag filled with biscuits. The *howri* visitors took tea breaks in groups. They sat on the paddy stalks and drank tea. Some dipped biscuits in tea and fed their children, while others sat under the banana trees and shared *beedi* (hand-rolled tobacco cigarettes) or rubbed tobacco leaves and lime on their palms and placed the mixture in their mouths.

Lunch was served at noon. Aka put out big pots of fish curry cooked with bamboo shoots, pork in dried yam leaves, steamed rice, boiled vegetables, and a dish of fermented fish chutney. As the *howri* visitors came to eat, they cut tender banana leaves and placed them on the ground, then took turns serving each other. Aka urged them on: "Khabi-khabi" (Eat-eat), and playfully apologized for not providing proper tables and chairs. They all laughed and one of the *howri* visitors responded that tables and chairs were not needed in the hills. They burst into laughter again. The older women served the young women, children, and teenage girls first. Once the younger ones finished their lunch, the children went off to sleep under the banana trees beside the granary, and the young women and teenage girls served the older women. In a corner, Aka arranged tobacco, areca nuts, lime, and *paan* leaves next to a large earthen pot, then called out, "Satini itu bhal pani ase" (Friends, this is safe drinking water), indicating that the water in the earthen pot had been boiled and was safe for drinking. That the water had been boiled and carried up to the field was an important show of concern for the health of the *howri* visitors. Aka ate after serving everyone else.

The care that was taken to serve the *howri* visitors indicated the importance of maintaining bonds among communities through nonmonetary forms of practice. However, *howri* was a common practice among cultivators, relations, and neighbors who possessed land. Adivasi *howri* participants such as Bina and her husband said they were casual workers on a tea plantation in Assam but had decided to come to the foothills in search of land and work. Initially, they worked as casual laborers on the tea plantations and as sharecroppers in the paddy fields. After Bina's family found land and settled down, they identified as farmers and started participating in *howri*.

In the upland Naga villages, residents explained that their villages were flooded with surplus workers from the foothills. These laborers worked in the rice fields and were paid cash or grains; reciprocal labor practices were absent. Residents from Bhandari told me that they did not practice *howri* with non-Nagas from the foothills. Instead they paid them, because they were strangers. Lotha Naga villagers engaged in a similar practice, called *aejonvoro*. In the Lotha Naga language, *aejon* means "to cultivate," and *voro* is a peer group. This was a group activity that was part of the agrarian structure in the hills, where they visited each other's farms to render reciprocal labor service. When I inquired how residents related to the *aejonvoro* practice, one of them responded, "It means that there are no strangers. They are people we know who are related to us. It might be our cousins, our in-laws, our clan members. They are part of the larger kin group, and they are close to us. We ask them personally to come and help us during the time when we need help."

Like the *howri* in Aka's *jhum* fields, residents in the upland village of Bhandari who observed *aejonvoro* also invited groups to work in their fields. But the practice involved the same clan and kin group in the village rather than being an inter-ethnic activity involving Adivasi and people outside the village. But in this case as well, feasting and drinking in the field was an important part of showing hospitality and gratitude. The owner of the rice field slaughtered pigs, offered liquor, and served grand meals. It appeared that *howri* had traces of *aejonvoro*, but there were distinctions and other meanings and attributes that produced new kinds of relationships. Unlike *aejonvoro*, where cultivators in the hills rendered reciprocal labor service as members of one clan or kinship group to reinforce their bond, *howri* practices involved forging new relationships, new kin networks and associations. In that context, *howri* transcended ethnic, religious, and social identities, unlike the upland Naga villages, which maintained kinship and ethnic boundaries.

Howri provided security, protection, and goodwill, but it was also paternalistic and unequal. In the past, while the scarcity of surplus labor in the hills might have supported practices like *aejonvoro*, it was different in the foothills, where there was abundant labor. Under such conditions, many residents who rendered *howri* services to each other were poor people who could not afford to pay wages or Adivasi families who held small patches of land. Friendships and solidarities, according to the families who practiced *howri*, were considered important relationships that provided security to those living in the foothills. Unlike the tight kin-based labor required for *jhum* in the hills of Nagaland, those engaged in *howri* were aware of the different systems of cultivation in use: settled farming, regimented plantation, and shifting cultivation. These foothill agricultural practices required a larger network of people than did the agriculture in upland Naga villages. *Howri* practices constituted a different kind of labor relationship that was neither coercive nor bound by rules of wage labor.

Yet not all relationships were defined with such fondness and attachment. *Howri* was limited to landowning families. Landless laborers who were adopted into households drove buses, worked in the coal mines, dug up the riverbed for sand and stones, cultivated the fields, and looked after the cattle.

ADOPTIONS

I met Gorib Maji, an Adivasi bus driver, at the Namsa *haat*. He shuttled traders to various weekly markets in the foothills. I learned that his official first name was Gobi but everyone called him Gorib (poor). He laughed and asked me to call him Gorib Maji. The bus owner, a Sikh man called Manu Singh, described Gorib as a *ghor manu* (family member). He stayed with Manu Singh's family in Sonari town, a coal mining hub. Gorib and Manu invited me inside their bus at the Namsa *haat*: "Oh didi, ahibi ahibi" (Sister, come in, come in). Gorib was loading the bus with stacks of broom, sacks of ginger, and huge rolls of *paan* leaves that the vendors from Assam had bought at the *haat*.

Earlier that morning, Gorib Maji ate a hard-boiled egg at the *haat* and shared his story:

> I started out as a handyman and then became a driver. I drive all the traders from Sonari to this market every week. I also carry the traders to three other weekly markets on the foothill border. The owners are Sardars from Sonari town. I grew up in their house. I carry around fifty passengers, but as you know, there is no limit to how many traders a bus can carry! A two-way ticket costs ten rupees per head. My parents are from Joboka, and they live in the Line Housing on the tea plantation. I have no education, and I never worked in the tea garden. All my five brothers work on the tea plantation, but I am the only one who drives a bus. I drive the bus since I learned how to drive, but there are many dangers, and there is no guarantee to my life driving this bus. I speak Hindi, Bagania, Bengali, Nagamese, and Assamese. But I do not know English, sister. I get five thousand rupees per month for my work.

Gorib Maji had been driving buses for twenty years for his adopted family. Business families or farmers from neighboring towns adopted Adivasis leaving the tea plantations. The landowners relocated them to their farmlands or kept them in the house as helpers. These practices were exploitative social relationships based in poverty. Descriptions of who constituted the family had little to do with flesh and blood. I do not suggest that people forgot about blood ties and ethnic relations, but they used the rhetoric of kinship to consolidate labor services, exchanges, and payments. Relations of patronage and familial ties are important, but the deepest desire of the

Gorib Maji at the Namtola *haat* (Nagaland)

landless family is to obtain land. Land is the anchor that makes one a member of the foothill community. Becoming a landowner is more than mere possession of a land title; it allows one to be recognized and be invited to meetings and gatherings, and to become part of a narrative on land ownership and claims to resources.

When I traveled to the vegetable farms in the upper elevations of Longtssori village in 2010, I met another Adivasi man, Birsa Munda, who had been adopted by a Naga farmer. He was living on the farm of his adopted Naga family and tilling the land. On the adjoining slopes a few kilometers away, his son was working as a daily wage laborer digging up the earth with a group of contractors looking for new coal mines. Birsa said he used to work as a casual laborer on a tea plantation before arriving in the foothills. "There is plenty of extra labor, so it is difficult to find work in the plains," he said when I asked him about life in the foothills. "There are lots of adjustments one has to make here."

During our conversation, he reminded me, "I am not a *halwa* [casual laborer]. I am a farmer." As Birsa and I conversed, it became clear that for him it was important to be recognized as a farmer. Many Adivasi households worked in farming, but they barely managed to make ends meet. Birsa's family lived in a mud house; they owned a few utensils, some threadbare bedclothes, and a mosquito net. It was a difficult life. In addition, the Assam police and Nagaland police who patrolled these areas harassed them routinely, making life unbearable at times. The Central Industrial Security Forces and the Central Reserve Police Force guarded the oil rigs and tea plantations not far from his house. Since personal conflicts between Naga and Assamese villages routinely escalated into violent encounters around the village, there was a constant presence of security forces. For instance, when cattle from an Assamese village went missing, the situation became a border issue if Naga villagers were held responsible for stealing the cattle. At such times, search parties often entered the farmland and threatened Birsa's family members, calling them thieves and suspicious characters because they were living on a Naga farm.

When I asked Birsa what happened during these conflicts, he explained that Adivasi families were stuck between the hills of Nagaland and the plains of Assam:

> When people from the plains come up here and attack the people and villages, we run down to the plains and take shelter. Escaping to the hills is dangerous; it means we are supporting the hill people and escaping there to take refuge. It creates problems for us to go down after that. The plains people target us and attack us when we go down. They provoke us, saying, "Of course, your villages are in the mountains." It is a difficult situation for both sides. When the deputy commissioner from Nagaland came here and asked me to be a watchman for the area, I refused the post. How can I guard the foothills or watch who takes timber, bamboo, and other resources from here? For the Adivasi people, there is no peace either on the hill or the plains.

Adivasi families like Birsa's had to choose between two distinct territorial areas—to be part of Nagaland or Assam. According to Article 371A, to possess land in Nagaland, one had to be a Naga or an indigenous inhabitant of Nagaland prior to the formation of the state in 1963. Many ethnic residents, including Nepalis, Kacharis, and Dimasas, were regarded as indigenous to Nagaland, but not the Adivasi. While in Assam, the Adivasi groups demanded scheduled tribe status to access concessions in education and employment, but these demands were opposed by various political fronts, and they continued to be attached to the tea plantations.

Birsa Munda reflected on his life in the foothills, saying that people were jealous of "hardworking people," referring to his qualities and desire to live a dignified life. He described his misfortune in a fragmented story about how he once possessed some land but lost it because people became *chuku laal* ("red eyed," or jealous). It seems he used to own land in an Assamese village, but people chased him away and occupied it. When he approached a shaman to figure out why this had happened to him, he discovered that someone had cast a black magic spell on his land. To demonstrate, Birsa took a tender banana leaf, poured some water over it, and placed his forefinger on the watery leaf. He began to draw on the watery leafy canvas, showing how the shaman had read his misfortune. He concluded, "It was a dangerous spell."[5]

Birsa Munda aspired to own land and become a farmer. Reflecting on the conversations about losing his land and the hardship that followed, he said, "We are here only in name. We are only *halwa*. We are landless cultivators and do sharecropping on the soil of others. We do not own any land here. Ideally, as *halwa* we should get rations, clothes, houses, and other necessities, but we do not get anything from the landowners."

Embracing an identity he had rejected at the beginning of our conversation, Birsa admitted that he was a *halwa*. He narrated his hardship and appeared resigned to the reality of the place. His everyday struggles revealed the violent history of the tea plantation in this part of the world and the new forms of extractive practices, such as coal mining and logging, that continued the dispossession and exploitation.

CONCLUSION

Land ownership and extractive practices such as coal mining are important for forging different kinds of friendship and kinship ties in the foothills. The desire to establish social connections is evident in power networks and alliances. Reciprocity and gestures of friendship are not limited to socially obligatory services such as *howri*, but are extended to connections that are fluid and exploitative in nature.

Networks of communications and bonding, including friendships and family ties, are forged in relation to land and resource extraction. Across the foothills, fantasies about oil, gas, and coal play a role in shaping the social world of the foothills, as militarization gives rise to violent encounters and aspirations. These disputes are situated among larger issues influencing the future of extractive resources in the region.

CHAPTER SIX

Carbon Fantasies and Aspirations

A LEM, a Konyak landowner overseeing a coal exploration project in Naginimora, explained the trade to me. "We cannot say who will find coal where and when," she said, and went on to detail the state's attempt to remove coal mining rights from the community. People were not sure how long this contestation between the coal mining communities and the government of Nagaland would go on. However, Alem was clear about one thing: people in Naginimora will oppose the state's takeover of the coal-producing land. After explaining how they had rejected a series of offers from the government regarding coal extraction activities, Alem concluded that they simply did not trust the state to take care of them: "They promise us employment and other benefits, but we have none of that. How can they take land from people who are uneducated and cannot read and write? We have to protect it, so we have all said no to the government's move."

Naga villages perceived oil and coal as resources that could radically transform their lives. In the Naga villages where the Oil and Natural Gas Corporation (ONGC) had explored for oil between 1973 and 1993, people fantasized about a prosperous carbon future and potential benefits from oil exploration. Oil and natural gas fantasies play a large role in the everyday desires and traumas in militarized communities engaged in resource extraction. Coal mining activities and fantasies about oil were shaping new alliances, power networks, and understandings about ownership and access to natural resources against the backdrop of a violent political reality. Irrespective of whether the coal mines and the oil exploration sites fell within the administrative jurisdiction of Assam or Nagaland, militarized power relations framed these carbon desires and fantasies.

During my fieldwork in the foothills of Nagaland, hydrocarbons generated exciting conversations about secure futures and fantasies of striking it rich. I wondered why, of all the natural resources in the foothills of Naga-land, people perceived coal and oil as symbols of power that could radically transform their lives.[1] In the foothill villages within the jurisdiction of Assam, hydrocarbon exploration was highly securitized. High concrete walls, barbed wire barricades around the townships, and security forces surrounding the rigs kept out the residents of neighboring villages. Although their lives were detached from the hydrocarbon business, accounts of secu-rity forces harassing, and in certain instances killing, local civilians were not uncommon.

Unlike Assam, where minerals including oil, natural gas, and coal belonged to the state, in the foothills of Nagaland, villages owned the coal mines and sold coal to traders and agents from Assam. The laws governing ownership and use of land in Nagaland were different. Article 371A of the Indian Constitution guaranteed rights over natural resources to Naga com-munities. In the last two decades, even the state government in Nagaland had attempted to interpret this provision in its favor, as control over natural resources became an important government agenda following the Indo-Naga cease-fire in 1997.[2] In this regard, resource fantasies became part of the everyday experience of people as well as government bodies.

The ONGC had drilled for oil and natural gas in Nagaland between 1973 and 1993, and remnants of its activity could be seen in the Naga villages. The ONGC started as a commission set up by the government of India in 1956 to conduct hydrocarbon exploration throughout the country, and eventually became a public sector undertaking under the Ministry of Petroleum and Natural Gas in India. In the hill state of Nagaland, ONGC began oil explora-tion in 1973 and continued until Naga political bodies and cultural associa-tions ended drilling in 1994 due to disagreements over royalties and land compensation. Until recently, abandoned wells and pipes marked this politi-cal history in the oil-producing Naga villages of the foothills.

Although people spoke about coal and oil fantasies, the relationship with coal and oil as physical matters was different. Unlike coal, which was acces-sible to the villages and could be extracted using rudimentary technology, oil was hidden and speculative. This very hidden quality of oil stirred people's imagination.

CARBON FANTASIES

What kinds of worlds appear in carbon fantasies? Yan lived in Champang ONGC, a village with abandoned oil wells in the foothills of Nagaland. I frequented this area during my fieldwork and was engrossed in the local debates over oil and coal exploration. Although some traders showed interest in rubber, tea, and timber, conversations about oil were always more exciting and attracted the attention of people in the village. One day not long after I had arrived in her village, Yan smiled mischievously and said, "There are stories going around in the village." It appeared that every time new people came to the village, there was speculation about which oil companies the visitor represented. There had been talk that I was in the village as a representative for an oil company.

In the middle of our conversation, Yan apologized for not offering tea and said that she could not enter the kitchen. While she was cooking dinner the previous evening, a snake had slithered into the kitchen. She screamed and struck the snake, but it disappeared through a hole near the hearth. "It is dead, I saw it! It is dead! But I am afraid that there is a dead snake in my kitchen," she told me. Then she laughed and said, "We can play a prank on my husband! He is always dreaming about oil. The entire village is always thinking about oil."

Her plan was to insert a metal pipe into the hole where the dead snake was trapped. Then she would invite her neighbors into the kitchen for tea and snacks and inform them that the ONGC had found oil in her kitchen. To prove there was oil, she would ask her neighbors to put their noses to the metal pipe and smell. "But it would smell of something rotting. How would that convince people?" I inquired, amazed at her imagination. "I will tell them that new oil smells like that," she said. "If you want to know anything about oil—news and rumors about oil in the village—talk to my husband. He knows everything about oil. He lives in an oil dreamland," Yan remarked.

Yan's fantasy about passing off the smell of the rotting snake as new oil condensed the hopes of people in the village. The ultimate goal of Naga villagers was to assert their ownership over the future of hydrocarbon explorations and extraction and prevent the state from taking away their land and resources. Such fantasies expressed identification with oil exploration and gave meaning to particular places. Yan, who spent a considerable amount of time in the kitchen, located her fantasy just under the hearth.

In the same spirit, when Yan spoke about her husband as someone who lived in an "oil dreamland," her tone held a hint of sarcasm. The knowledge she attributed to her husband was not exclusive to him, however. Rumors about oil agents visiting the village, state debates over oil exploration, and reported sightings of oil leaks in abandoned drilling sites shaped all kinds of relationships, from the hostile debates between Naga citizens and the Nagaland state all the way to marital bonds. These fantasies were never focused on the individual alone; they also drew in neighbors, landscape, ecology, family, and state officials.

Yan's carbon fantasy that particular day started with her inability to enter the kitchen, her hatred of snakes, and her frustration over her husband's inability to find employment. But she skillfully included her husband, the neighbors, and eventually the entire village. She ended her fantasy about the smell of new oil by saying, "My husband lives in an oil dream world," and quickly attempted to disassociate herself from the fantasy world. Having spent time with Yan at the *haat* and her *jhum* fields, I noticed that she often made a conscious effort to separate her life from that of her husband, who was living in an "oil dream world." For some reason, the attribute of being a dreamer in this village, according to Yan, was closely tied to thinking about extraction and grounded on oil.

Yan described her husband as a dreamer who constantly imagined new business ventures that never took off. But among his plans, the dominant ones involved oil. Yan's family lived in a bamboo hut, like most of the villagers. Although the village lacked electricity, schools, or primary health care services, the government had set up a bank and a security checkpoint there. During my fieldwork between 2006 and 2011, insurgents and security forces alike used this village as a transit camp to monitor the foothills. The poverty and hardship in the village were stark, as was the everyday militarization of the place, and these social realities were entangled with speculation related to hydrocarbon exploration.

The militarization of the foothills and the harsh living conditions faced every day by subsistence cultivators like Yan's family were apparent. The governments of Nagaland and Assam declared 2009 and 2010 as periods of drought. During this time, small farmers and subsistence cultivators were not able to plant crops and depended on subsidies from the state. Because of a dysfunctional administration and corruption, much of the relief budget went missing. This meant that many poor families, like Yan's, received no financial help. For two consecutive years, her family struggled to secure water as wells and streams dried up. The image of the frontier as a land with abundant resources began to tempt its own inhabitants. Every time stories about new companies, agents from Assam, and potential coal mines started to circulate, it seemed as though their poverty and hardship might be over.

There were plenty of conversations about the future of oil exploration in the foothill villages of Nagaland. Naga villages had the power to decide what activities took place on their lands. Carbon fantasy, in this context, captures two important aspects. First, instead of perceiving fantasy as wild and unrealistic escapist daydreaming, fantasies can be understood as the magical creativity and pleasure conjured in constructing the resource frontier. Second, fantasy captures the moment when material realities of place and imagination unrestrained by authority and logic appear as spectacles of opportunity, tragedy, and violence. In this social milieu, carbon fantasies are reiterated and kept alive as bureaucrats, reporters, traders, security forces, and insurgents constantly reconfigure the internal mechanisms of power and networks in anticipation of resuming oil and gas exploration in Nagaland.

OILY LANDSCAPE

Wealthy Nagas bought vast plots of land in the coal mining sites, and in villages and the adjoining hills with abandoned oil wells, as investments for the future. According to official reports, the ONGC abruptly abandoned twenty-nine oil wells and two gas points in 1994.[3] The Minorities at Risk Project's "Chronology for Nagas in India" outlines the ONGC's exploration activities in 1994:

> April 30, 1994: India's state-owned Oil and Natural Gas Corporation (ONGC) halted exploration in Nagaland after a separatist Naga group bombed its offices and warned it to stop all work.

> May 31, 1994: Naga students call for a day-long strike in protest of the Indian state-owned Oil and Natural Gas Corporation (ONGC) warning that oil exploration would not be allowed unless all their demands were met.

> June 1, 1994: The Naga Student's Federation (NSF) was protesting the Oil and Natural Gas Corporation drilling of oil without the consent of the local inhabitants of Nagaland. The strike brought to a halt traffic on the Dimapur-Imphal national highway. In Kohima, all shops, business establishments and educational institutions remained closed. Central and state government offices were open with thin attendance.[4]

The chronology also captured violent clashes that took place between Naga insurgent groups and the government of India between 1990 and 1999. Now, abandoned oil wells and pipelines stand as markers of the political history in the oil-producing Naga villages of Nagaland.

When I followed residents to an abandoned ONGC site in the foothills, they told me that it had been eerie to enter these sites soon after operations shut down. The villagers found wooden beds, bedclothes, and books in the trailers. They dug out small oil pipelines and disassembled machines to sell for scrap metal. But the giant oil pipelines remained fixed in the ground.

On the outskirts of the village, the oily giant pipelines stood on the grassy patches of the village land. These abandoned oil pipelines and metals protruding from the earth continued to dribble dark crude oil.[5] On the outskirts of the village, a concrete path constructed by the ONGC in the 1970s led to abandoned pipes covered with wild grass, nettles, ferns, and creepers. This oily field soaked in crude oil and covered with green grass served as village grazing land. Cattle roamed here, selecting green patches of grass and avoiding the ponds of crude oil. Beyond the oily grassland, forest had reclaimed the ONGC office.

Leftovers and residue from the oil sites found their way into the Naga villages. Scraps from the oil sites in Nagaland captured people's interest. A pair of gloves, a piece of pipe, tools and metal pieces from the oil rigs, or dogs with dripping black paws started conversations about a carbon future. When I met Mary from Champang ONGC, she wore thick white gloves to protect her hands from cuts and bruises in the rice field. The gloves belonged to her brother, who was a contract worker at an ONGC drilling site. I met Naga landowners who were excited about the possibilities of oil exploration. Landowners like Mr. Kithan, on whose land the ONGC had once drilled for oil, were most enthusiastic: "I am watching over the oil reserve. That time there were ten wells, now there is oil in eight wells in this foothill. Here, there are sixteen cultivators on whose land we have found crude oil. This is the first place in Nagaland where we have found oil. ONGC was exploring for oil here but due to the land compensation issues and other conflicts that led to problems, it has stopped. . . . We found oil here in 1983; it was declared. Until 1993, ONGC worked here, but it was stopped. This was the first place in Nagaland where we found oil."

Mr. Kithan's views were reminiscent of many wealthy Nagas who owned land adjacent to oil and gas sites in the foothills. But these aspirations were also prevalent among cultivators who possessed small patches of land. They were not part of the powerful associations of landowners at the forefront of asserting rights in the Naga villages. Yet the aspirations of subsistence cultivators struggling to feed their families were also significant. Despite the environmental damage still visible in the Naga coal mining villages, the promise of a carbon future informed the politics on the ground.

Aspirations around coal surfaced in different ways from those related to oil. Digging for coal took different forms. Ranging from small artisanal coal mines to large-scale operations, coal mines dotted the landscape of many Naga villages in the foothills. Coal mining opened up a world of alliances and trade and created livelihood opportunities in several villages. Many landowners leased their land to coal traders from Assam in the hope of making a profit and attaining a better life. The visible wealth of coal trading families stimulated frequent conversations about profits from mining.

As competing groups rushed to the coal mines to claim their share of the wealth, coal became a source of conflict between villagers and authorities (both state and nonstate parties) in Nagaland. Coal exemplified the circuitous networks of authority and control involving individuals, kin groups, village units, and state government in Nagaland. The coal operations also revealed that tribal societies were not homogenous and egalitarian, as issues of class, social hierarchy, and power relations surfaced in these villages. It is true that coal mines in Naga villages followed customary law, which asserted that people had rights to land and natural resources. But the monetary profits belonged to individual landowners and not the collective community. Individual households kept the profits for themselves and invested in exploring for new coal deposits.

"I am king of the jungle!" This was Toshi's favorite line. He was a landowner and a coal trader in Nagaland. His village had been established in 1989 but was not officially recognized by Nagaland until 1994. This meant the village was listed in the official state gazette much later. State development funds and other subsidies started arriving in 1994. On a humid October afternoon in 2010, I followed a group of traders from Anaki Yimsen to visit some coal mining sites outside the village. Toshi grabbed a bottle of rum from his house and joined us. Along with a small group of coal agents from Assam and some villagers, we followed Toshi to his coal exploration site. His brother Meren, a former soldier in a Naga insurgent group who was supervising the project, told us they had been camping there for three months. Meren was living with the workers in a large dormitory constructed with bamboo and large plastic sheets.

"I am king of the jungle!" Toshi exclaimed again. Then he asked me in Nagamese, "Itu teh interest keleh?" (Why are you interested in these things?). He watched the activity in the surrounding hills as I explained what I was doing there. Land excavators broke down the hills and dug ever deeper in search of coal. As we watched from the dormitory, Toshi announced that he was going to stage a play. His theatrical creativity appeared to be inspired by the afternoon sun and the bottle of rum he had finished with his workers. He ordered one of the workers to stand beside him and enact the role of a security guard. The worker picked up a stick, held it as a gun and pretended to protect Toshi. "The king needs protection," Toshi declared. He demanded that I take pictures of him while he struck poses.

As the landowner, he possessed the coal mines, trees, and everything on the mountain, and his brother Meren managed the operations and the workers. Together they wielded immense power. In 2010, Toshi leased a large mountainous tract to a business enterprise in Assam. The agreement was simple: he provided the land, and the company invested money and manpower. If the company discovered coal, Toshi would get a percentage of the profits, in addition to rent on the land. Hundreds of such deals had been struck between Naga villagers and coal traders from Assam. On our way to the second coal camp, we began to climb a higher mountain. "We are in Mokokchung district now. But when we cross over to the next mountain and enter the second coal camp, we will be in Mon district," Toshi said. The mountains were contiguous, but landowners were mindful of their operational zones and boundaries, as conflicts were frequent among landowners engaged in mining.

In addition, there were also routine conflicts between tribes over the district boundary, since the eleven districts in Nagaland were divided among tribal groups. "It is peaceful now, but what can we say about tomorrow?" a resident from Toshi's village commented. Anaki Yimsen was involved in a boundary dispute with a neighboring Naga village, and the residents described burning rice fields and physical attacks in the recent past. Therefore, the village was on alert and had planned ingenious ways to protect its residents. The pastor of a Naga village adjacent to Toshi's coal camp said the church members and village council had come up with an innovative way to protect their village documents. They saved important documents and records on a flash drive and hard disks. The village chairman told me that the flash drive was kept in a safe location for protection: "If by chance something happens to our village tomorrow, if there is a border dispute, we cannot take our village records and escape. But we can take the flash drive and the disks."

When we arrived at the second coal site, a Konyak man in his twenties welcomed us as honored guests. We learned that he was the manager overseeing the project. He invited us in Nagamese—"Ahibi! Ahibi!" (Come, come)—and hurried out of his bamboo camp to shake our hands. Now we were officially in Mon district, and the land we were standing on belonged to a Konyak businessman. The young manager said the landowner had a budget of ₹1 crore (10 million rupees, approximately US$115,800) to explore for coal in the mountains. With the noise of the yellow Caterpillar excavators mowing down trees, flattening *jhum* fields, and breaking down the mountains behind, conversations about the ongoing coal exploration started among the coal traders. I learned that the Konyak landowner had spent 46 lakh rupees (approximately US$53,000), but the team had yet to find coal. "We have a November deadline. If we do not find coal, we will move to the next mountain," the young manager noted. "Ah, what a loss—that landowner is never going to recover," Toshi said, shaking his head in dismay. They turned to me and explained, "Coal is a gamble. You need luck. Either you are lucky or you run up a huge loss." We were in Longleng district visiting the coal mines, but Toshi's reference to the site as part of Mon district seemed to indicate the continuation of coal mining activities in the area.

GENDERED DESIRES

As we began to trek down the mountain, Toshi followed us. He appeared excited and started to walk beside me. "Here, let me help you." He smiled at me, came closer, and started to rub my back. I refused his help and began to walk faster. The extractive economy is predominantly masculine. Coal extraction created exclusive bodies of male traders, politicians, private armies, and insurgents, giving rise to an extractive economic culture of "hypermasculine militarism" (Mbembe 2005: 148).

"You will not understand what we are talking about. It is man talk," said the men, dismissing the women who listened to their stories about profit and coal. Male workers at the coal mines condescendingly asked me, "Why are you working so hard?" when I inquired about their trade. When men gathered around fireplaces in the kitchen, drinking and plotting how to make money and dig up more mountains, they ordered the women to cook for them, clean their muddy shoes, and heat water for their baths. These masculine discourses played out in households every day across the villages of the foothills. In addition, the sight of women in the coal camps transformed the behavior of the coal traders. When I visited coal mines, workers and traders often teased one another and commented, "Smile for her, smile at her. It will be a sleepless night for you today." Sexual jokes, erotic connotations, and marriage proposals entered the conversations. The mining sites were deeply gendered.

The exclusion of Naga women from positions of power and decision-making forums like the tribal councils was based on Naga customary law, which prohibited women from making decisions or inheriting ancestral land and property. In oil speculation and coal mining, new forms of gender inequality, violence, and dispossession were being produced. Women were excluded from the emerging new economy centered on land and extraction of natural resources, but their lives were deeply affected by the mining. As the scramble for coal sites intensified, the desire for profit increasingly threatened women's spaces in the foothills. The small patches of *jhum* fields that sustained women-headed households and allowed them to sell produce from the field in the *haats*, as well as reciprocal labor services like *howri*, would disappear. Wealthy Nagas arrived in the villages to buy up land in anticipation of opening new coal mines or conducting oil exploration.

In Naga villages, households engaged in subsistence agriculture improvised to generate income from the land because it was impossible to sustain the family solely from farming or coal mining. During the lean season, households leased land to neighbors or itinerant groups from Assam who would otherwise be denied access to forests and community lands in Nagaland. During the coal season, some of the women worked as cooks or helpers in the coal mines. They were also active in the *jhum* fields. They decided what crops to plant and which produce to sell in the *haats*, but decisions about buying and selling land or coal mining operations were made by the male members of the family. As coal mining expanded, new kinds of power relations and gendered inequalities emerged.

Gendered Accounts of Coal

I met Nate in a Naga coal mining village. She hosted me and took me to visit the coal mine when I arrived in her village around November, the high season for coal mining. Her manager in the coal mine updated her about the work and the rations required for the workers. One morning, Nate and I went out to inspect the coal depots and check on workers loading the trucks with coal. The workers, drivers, and trucks came from Assam, and it was Nate's responsibility to ensure that goods and workers left the village safely. She told me that Naga insurgents frequented the coal mines and the village, but they did not cause harm because they were paid to provide security and maintain order. A babysitter carried her youngest child, who was less than a year old, and brought the baby to Nate at regular intervals to be breastfed. Nate controlled the coal trade and managed the household. She was the only female coal trader in the village.

Nate explained that she had married into a powerful family of landowners in the village, and her husband soon became involved in the coal business. They became wealthy and powerful after rich coal deposits were found on their family land, but the mining operations also brought enemies and jealousy. Her husband was the eldest child, and responsibility for the family's fortune rested with him. Nate managed the accounts and helped her husband run the coal operations. After her husband was killed in an accident, Nate had to take care of the children. She wanted to make sure the activity in the coal mines was not disrupted, since this was the family's sole income. As a woman, she could not inherit her husband's land or have custody of her children, so she married her husband's brother to retain control of the family mines. Soon thereafter she took over management of the coal mines, since she was experienced and the family trusted her. Even though she did not own the land, she was the authority figure and controlled the workers. Nate accepted that the land belonged to the men in the family and that her sons were the rightful owners of their vast property.

Nate's life underlined widely held conceptions about land, ownership, and the role of the family in Naga society. It was not women's labor that threatened Naga cultural and family values but rather female inheritance. Naga masculinity was based not on providing for the family but on the custom that only males inherited land and property. When I visited the coal mines with Nate, the workers referred to her as the "man" in the family, clearly indicating that she was the boss. The manager of the coal mines often said, "Tai toh maiki holebi mota manuh nesina kam kore" (She is a woman, but she works like a man), thus suggesting that unlike other women, who were seen as sexual objects in the hypermasculine coal mines, Nate demonstrated masculine characteristics. Yet she did not threaten the patriarchal system because she worked for the benefit of her husband's family while also carrying out her responsibilities as a caregiver at home.

A similar case involves Jyoti, whose husband was an important coal trader from Assam who was well connected in the Naga mining villages. He had been adopted by a powerful coal mining family in a Naga village, and Jyoti said that she was treated as a daughter-in-law when she visited Naga villages. The Naga landowners would tell her, "Toi toh bosti laga swali" (You are the daughter of the village). When Jyoti's husband passed away, the landowners and traders gradually stopped discussing coal matters with her, although she had inherited the coal trucks and the coal depot outside her house.

Instead, the villagers gravitated toward her brother-in-law, who ran a grocery shop. When Jyoti asked me to accompany her to the village, I obliged. During our two visits, the landowners' families welcomed us warmly and were affectionate toward Jyoti. They slaughtered a chicken in her honor, heated water for her bath, and offered her the best bed in the house. However, when we sat around the fire and Jyoti expressed interest in taking up her late husband's business, the landowners refused to talk about the coal mining operation.

Unlike the joking and camaraderie between male coal traders from Assam and Naga landowners, my visits with Jyoti were somber affairs. The discussions were polite, the dinner was quiet, and it was the womenfolk who kept us company, asking about our travels and describing their lives in the village. Eventually, Jyoti's brother-in-law inherited the business contacts of his late brother, and Jyoti gave up the coal trade. This was neither unusual nor new. It was typical of alliances and practices in coal trading. The masculine display of power and authority in the coal mines and the fixed gender roles grew out of the extractive activities. These practices of gendered exclusion became the basis for new configurations of power and masculinity in Naga coal mining villages as well as coal trading hubs in Assam.

Yet the gendered aspect of coal was not the only distinctive factor. When I returned to the village in March, it was the peak of coal activity, and the traders were upset. Nagaland's minister of geology and mining had ordered the village to close down coal mining and trading until it had paid the coal tax. In 2014, Nagaland had amended the Coal Act of 2006 to regulate mining activities in Nagaland.[6] According to the *Nagaland Gazette*, the official state bulletin, the Nagaland Coal Policy (Amendment) of 2014 was put in place as an urgent measure. Section 2.2, "Present Status of Extraction of Coal," states the government's position:

> 2.2 However, Un-planned extractions of coal are still being carried out by private parties/landowners at many places, especially in Mokokchung, Wokha, Dimapur, Mon, Longleng, and Peren districts. These rampant and illegal mining activities have resulted in various types of accidents, health hazards, ecological and environmental degradation besides loss of coal resource and leakages through process exits have caused substantial State Revenue loss. Therefore, to adopt a systematic development of coal resources available in the State, and their optimum utilization in industries and for other purposes; the Nagaland Coal Policy & Rules (Amendment) 2014 is being adopted.[7]

The description of legal coal mining operations was unclear. The government's main concern, aside from ecological hazards, was the loss of state revenue. When I visited the Directorate of Geology and Mining in Dimapur (Nagaland), a Naga geologist explained the position of the government: "The reason why the government wants to intervene and regulate coal mining activities is for revenue purposes." The relationship between the villages and the government was acrimonious. Yanger, a coal trader from a Naga village, described the first official visit of the geology and mining minister to his village in 2010. The residents questioned the minister about his intentions because they were aware that his arrival meant trouble. In addition to the land tax, the insurgent national tax, and the village council tax, the coal traders were worried about a new coal tax payable to the state.

This new coal tax significantly disrupted existing practices of taxation and tributes in the Naga coal mining villages because it meant reworking the existing coal payments. When a landowner discovered coal, he leased out his land part by part. The lessee paid a land tax to the landowner. But when business groups from Assam financed coal operations, they signed a memorandum of understanding and paid a coal royalty to the landowner in addition to the land tax. In addition, there were other calculations to consider. A truckload of coal is sold for ₹5,000–5,500 (approximately US$70). The Naga landowner was paid ₹1,000 (US$20) for every truckload of coal. On the long journey from the Naga village to Assam, traders stored the coal in temporary depots. Since the foothills were covered with vast tea plantations, the coal depots were often on tea plantation land. These detailed coal taxes and anxieties about the presence of the state meant that Naga coal mining villages had come up with a system for carrying out the mining activities. This was also the case when it came to locating new coal mines.

FEELING COAL

As I walked around the coal camp taking in the sights, I saw that Toshi was "king of the jungle." He was sitting on a hillock, drinking rum and eating meat from a bamboo bowl. The coal venture on his mountain was not going well; no coal deposits had been found. To make up for lost time, Toshi and his son, who was home for winter break, had leased a coal mine from a landowner in the village. A trader explained why Toshi's venture had failed: "See, the mining company [working on Toshi's land] has no idea about coal. They have no idea how to mine or carry out coal exploration." It was not enough to dig up the entire mountain looking for coal.

What did it mean to have an idea about coal? Later that day, I crawled into a narrow mine, following the voices of two young men working inside. "Come on, you can enter. We are down here," the miners howled from the depths of the cave when I asked them if I could come in and talk. The mine was narrow, dark, and humid. The ground emitted a strong odor and vapors that burned the eyes, nose, and throat. We sat down to talk about coal, breathing in the pungent gas and surrounded by black walls of coal. The workers carried candles and flashlights as they dug farther into the vein. We

Inside a coal mine (Nagaland)

seemed to be at a dead end, but the two young seasonal workers contemplated in which direction to dig.

Under the ground, inside the serpentine openings, people developed ideas about coal. They felt the coal nugget with their hands and studied the black lines marked on the walls of the freshly dug caves. Following the coal belts, they dug deeper into the earth searching for large deposits. Inside the mine shaft, I sat on a pile of coal dust and watched them. A flickering candle stuck on the wall made us appear like ghostly figures. "We work like this, and we get paid two hundred rupees as daily wage. We pray and enter these mines every day," the workers said. "How do you work in the dark? How do you know where there is coal?" I asked. "We follow the coal veins. That is why these tunnels are not straight. We keep digging wherever we see coal. In some cases, the earth starts to cave in, so we try and work as fast as we can and excavate as much coal as we can carry," they said. They held the candle close to the wall to illuminate a dead-end tunnel I had not noticed.

One of the workers said, "Look here! Since the coal vein ended here, we stopped digging here. We abandoned this tunnel and turned our attention to the next coal line and then the next one." They felt coal through their senses. First it was visual. They paid attention to the color of the soil. Then it was a smell. The gray soil began to emit fumes. As they began to dig in those marked spots, they found lumps of coal or coal dust. The smell and texture of the soil changed in front of their eyes. In contrast to land excavation machines digging randomly, the workers believed that human intervention was essential to discover coal. Every time a new patch of earth was opened, they were needed to read the coal signs.

The adjacent coal mine was leased out to two college students. They were also young men from the village, but unlike daily wage laborers or seasonal workers, they had leased a mine and were working the site with their miners. Since the beginning of coal season in November, they had managed to get fifty truckloads of coal from their mines. They planned to continue working until April. Describing the quality of their coal, they said, "We mainly found coal dust, but it is the best coal dust," emphasizing their remarkable discovery. There was no expert here, but the young men confidently claimed that they had also found coal nuggets and categorized them as "good-quality coal." Altogether there were five or six coal mines leased to various parties. It was a diverse group: college students from small towns looking to earn pocket money, high school students from the neighboring Naga villages, Adivasi teenagers from the tea plantations, Ahom boys, and Bihari and Bengali laborers from Assam. As the workers carried the dust and nuggets out of the dark openings, a new batch of workers filled up the trucks lined up ready to head down to Assam.

CONCLUSION

Extractive regimes and speculation around both oil and coal in the Naga villages generated fantasies and aspirations. Carbon fantasies and aspirations permeated social and political boundaries and produced new alliances and configurations of power. Insurgents, politicians, wealthy landowners, and cultivators alike aspired to participate in the future of oil, just as they continued to dig out coal and expand mining projects in their villages. However, beneath the excitement and confident competition for a prosperous future, militarization and violence shaped the political reality of the foothills. Thus, the flow of coal and conversations about the future of oil took place within new regimes of control and order that comprised politicians, insurgents, landowners, and investors from Assam and beyond.

CHAPTER SEVEN

Carbon Citizenship

T HE states of Assam and Nagaland function within a political structure of militarization and resource extraction legitimized by the Indian state. For the greater part of the twentieth century, political movements in Assam centered on the largest extractive economic regimes, particularly oil. In Assam, the oldest oil-producing state in India, hydrocarbon exploration is a well-organized economic sector. The Assam unit of the Oil and Natural Gas Corporation (ONGC) extracts 1.2 million tons annually. The hydrocarbon reserve in the state is estimated at 1.3 billion tons of crude oil and 156 billion cubic meters of natural gas (Ministry of Petroleum and Natural Gas 2016).

The processes of oil exploration in the foothills of Assam illustrate the formation of a carbon citizenship regime in Northeast India where knowledge, procedures, and legal right of its subjects are drawn from the extractive resource regime. The dynamics and processes of exploration and extraction are deeply shaped by a security prism. Unlike the conflict over land ownership and natural resources between the Naga coal mining villages and the Nagaland state, for villages and towns in the foothills in Assam, the framework of this securitized carbon citizenship produced in the oil towns and drilling sites significantly influences everyday lives and politics. The experiences of residents across towns and villages in the oil-producing areas of Assam reflect everyday practices of extraction, citizenship, and militarism.

Nagaland and Assam secured their own foothill borders, but it was different for the Indian armed forces, often called the "central forces." Reporting directly to New Delhi, it is the central forces that protect oil and gas sites and townships. They perceive the foothills and the region predominately as an unsafe and dangerous place where extremism and terrorism are rampant (Borthakur 2015). Oil officials share similar concerns. What seemed

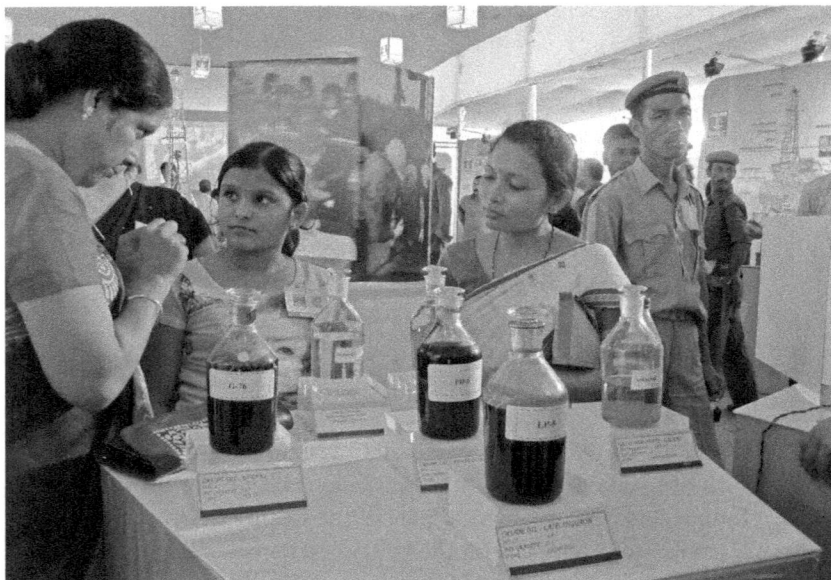

Crude oil at the ONGC carnival (Assam)

like disparate accounts of technical matters, culture, and regulations that ONGC technical teams shared with me became a disturbing memoir of carbon citizenship practices. Oil exploratory operations and counterinsurgency in the foothills of Assam significantly framed each other. These two activities were monitored by the Indian state and gave rise to a process that determines reliable, good citizens so as to support the hydrocarbon operations and administration of the oil and gas drilling operations in the foothills. Oil officials and the central forces together, I observed, invented a framework to define the *bhal manu* (good person).

Among the governments that residents categorized as Assam, Nagaland, and India, it was eventually the Indian state that emerged as the most powerful. Through extraconstitutional regulations like the Armed Forces Special Powers Act (1958) and the Disturbed Areas Act (1955), which were imposed across Northeast India, including the foothills of Assam and Nagaland, Indian security forces applied a logic of military surveillance to profile people and protect oil and gas exploration. High concrete walls, barbed wire barricades around oil townships, and security guards at the rigs kept local people at bay from the oil and gas gathering sites. Unlike Naga coal mining villages, where landowners, coal traders, workers, and neighbors came together, ONGC sites were reserved for experts and the technical staff.

As in any large corporation, ONGC employees were attached to various divisions and often did not know their colleagues. For example, the team working with the drilling units had different schedules from the soil and mineral testing units in the laboratories. The duties of each division were carried out autonomously, and each division operated efficiently so that the system functioned smoothly. Unlike the residents, who mingled and shared their accounts with neighbors and friends (as well as researchers like me), ONGC executives and scientists were extremely cautious and steered clear of conversations on hydrocarbon topics, even during social gatherings. Oil, natural gas, and coal were sensitive issues. Insurgent groups in Assam, such as the United Liberation Front of Assam, which was fighting for a sovereign nation, had a long history of blowing up oil pipelines as a political protest against the Indian state. Oil was extracted from highly visible spaces around human dwellings across the foothills, yet the lives of the oil world and the villages were often disengaged from each other.

THE OIL CARNIVAL

"This is the first of its kind!" announced a representative of the ONGC Corporate Social Responsibility office, explaining that its carnival was intended to create public awareness about the benefits of oil and natural gas exploration. The Gyan Kumbh Mela (Mega Knowledge Carnival) was held in Sibsagar town on August 10, 2009. Inside the tent, samples of rocks and soil labeled with names of villages and towns, including Gelakey, Anaki, and Namsang, were categorized by their topology. A neatly sealed display showed several kinds of soil taken from the foothills in a transparent glass tube. A big signboard next to the glass tube announced, "A Look through Rock Sequences Penetrated in Walls (Upper Assam)." These rocks and soil samples had been collected from various sites where the ONGC had discovered oil and natural gas. Consisting of soil and dust particles in a glass tube, the rock sequence was divided into three sections showing the character, color, and texture of the soil, along with its depth in meters. The rock exhibit featured soil types named after indigenous communities living in the foothills.

The carnival displayed impressive technological and scientific expertise. Oil technicians and engineers walked around the carnival tent offering to talk with the audience—hordes of schoolchildren, college students, housewives, and the wider public—about the ONGC's activities. I approached an ONGC engineer standing next to a giant map and asked, "Where is oil found in the northeast region?" He gave me a quick geography lesson. The area was bounded on the northwest by the Himalayas and on the southeast by the

Naga Hills. The fault lines in Northeast India ran along the foothill areas. As I pointed to a large map titled "Naga Thrust," the engineer explained, "Thrusts are several faults." Two kinds of rocks were found in the foothill borders: the layers of rocks that held the oil, known as reservoir rock, and trap rock, which prevented the oil from getting out. "The purity of the oil depends on the genesis of the raw material." As the geologist continued to describe the Naga Thrust, he highlighted rocks and soils, oblivious to the ethnic names and categories assigned to each piece of carbon sample, soil type, and oil well displayed at the carnival.

"We look at the structure of the rock and then think about oil," the ONGC geologist explained. The physical composition of oil is not the same in all oil wells. Some oil is lighter and some heavier. Some is "waxy," meaning it contains more coal tar. Then he declared, like an oil aficionado, "Every product from oil is useful." Later that day, a few ONGC officials invited me for lunch at the office canteen. They smiled and remarked that oil companies are like gamblers. Investing did not necessarily mean a guaranteed return. For this reason, the ONGC preferred to rent drilling machinery. The organization applied for tenders from private companies like Halliburton, Schlumberger, and Weatherford, which had headquarters in Europe and North America. These machines arrived in Assam disassembled, and technicians reassembled them at the oil exploration sites. "The technology in India for underground horizontal drilling [allows us to drill to] . . . one kilometer, but we have to learn the technology from Russia," the ONGC geologists and engineers explained. Russian oil technology allowed horizontal drilling for oil to a depth of ten kilometers. The geologists thought that if they could acquire this technical knowledge, they could drill in the foothills of Nagaland.

"How do you detect oil?" I inquired, and one of the geologists explained the procedure. Field Digitization Unit monitoring machines are used to detect oil. Across the foothills, hundreds of magnetic coils attached to wires were inserted into the earth, and mechanical energy was sent down to pick up oil signals. The mechanical energy was converted into electrical energy as the digitized system amplified the system through the magnetic coils. When oil was found, the area became a "zone of interest," and a team of thirty to fifty technicians was dispatched to the site and an average of one thousand laborers hired.

However, the ONGC categorically separated the violence and political realities of the place. To them, the foothills were defined by technical terms like operation sites and zones of interest. Especially, the zone of interest was a technical definition to determine the prospective oil sites as they carried out exploration in the foothills. Any questions about resource conflicts or the insurgency or security issues resulted in silence. Cautious about discussing "political matters," one of the geologists said that ONGC did not want to politicize "technical" issues.

Conversations with the geologists and engineers at the oil carnival painted hydrocarbon exploration as a depoliticized project. The ONGC's classification of the militarized landscape as a zone of interest obliterated the people and their histories, transforming the place into a hydrocarbon site. Its value was limited to its hydrocarbon composition. Oil activity operated as an "anti-politics machine" in the foothills, to use a term that James Ferguson applied to the way political questions (in the development apparatus) are reframed as technical questions requiring a technocratic intervention. This takes place, according to Ferguson, through "reposing political questions of land, resources, jobs, or wages as technical 'problems' responsive to the technical 'development intervention'" (1994: 270).

The distinction between expert knowledge and the political reality of the people gives rise to a militarized emphasis on value and security. For those in the oil townships and exploration sites, the world outside the walls is insecure and violent. For people in the villages and towns, the barricaded camps are dangerous spaces. The absence of spaces shared between those inside the oil barricades (rigs, townships, and offices) and those outside (villages, towns, social institutions) produced wide-ranging perceptions of local communities as dangerous and untrustworthy and led to violence at the hands of security forces in the foothills.

THE ONGC OFFICE

The ONGC regional office is located in Nazira town in Sibsagar district, which is also regarded as the intellectual hub of separatist resistance in contemporary Assam. Nazira is steeped in history. This town is around fourteen kilometers from Sibsagar, the heritage town that was the seat of the Ahom dynasty, which ruled Upper Assam for six hundred years.[1] Important historical Ahom sites scattered across this town are maintained by the Archaeological Survey of India and frequented by tourists and locals alike. Like the security barracks and tea plantations, the ONGC oil township in Nazira is fortified and fenced off from civilian areas. Schools, banks, and recreational and sports clubs catering to oil employees and their families are located inside the township. The township is built on the former site of the Assam Tea Company. Thus, some oil offices, the hospital, clubs, and guesthouses are in colonial-era buildings. The narrow one-lane paved road that connects Nazira and Sibsagar town is usually busy. Big oil trucks loaded with pipelines and buses carrying people with "On ONGC Duty" emblazoned on their windshields flew past me at regular intervals as I traveled here.

"Assam Asset Celebrates 15th August. Million Barrels, Billion Smiles," proclaimed an ONGC billboard outside the Nazira township. At the entrance, I was given an entry pass and required to sign a register: time, date, name, purpose, address, and phone number. At the second gate, security guards checked my bags and insisted that I leave my mobile phone, camera, and tape recorder at the gate. At the third gate, my bags were checked once again, and I was asked to show my entry pass. At the last gate, policemen asked me to again sign a register. As I entered a long corridor with several doors, I was directed toward the Office of Corporate Social Responsibility. The ONGC office in Nazira was a modern fortress with continuous heightened security procedures.

Inside the office, the Corporate Social Responsibility (CSR) officer offered me a seat. Before I could say anything, he enthusiastically said, "The ONGC is extremely open about everything we do, and you are welcome to visit us anytime!" The CSR officer, seemingly unaware of the extensive security that I had just experienced, continued to praise the "friendly spirit" of the organization. When I showed him my fieldwork letter and offered details about my research project, he quickly cut me off: "ONGC is open about our work." He explained that I should "feel free" to ask him "anything." But when I started to ask about ONGC's activities, he said he was bound by regulation not to discuss ONGC activities in the foothills, or for that matter any ONGC operations in the "region," meaning Northeast India. Instead, he turned to an officer sitting beside me and began to talk office politics.

"My work is not appreciated," the CSR officer explained, and said that his transfer was due. "When I was posted in Assam, my colleagues told me that it was a punishment posting. I was working in Delhi for the most difficult boss, but it was because of my work that I survived. Since I am in Assam now, I want to do what I think should work here," he continued. However, since arriving at the ONGC office in Nazira a few years ago, he had been unable to set up a structure that would allow workers to diligently carry out their duties in the office. The CSR officer blamed it on the difference in work ethics between the New Delhi office and the Nazira unit. He expressed his displeasure that local employees took long lunch breaks. When he asked about scheduled time for lunch breaks, no one in the office, including the office clerks, could tell him when lunchtime started. Almost every office matter was hampered by differences in work ethics between Delhi and Assam. However, problems outside the ONGC office and beyond the township were even larger.

The CSR officer conceded that there were problems in the exploration areas in the foothills, but he described them as cultural problems between ONGC employees and the public, meaning residents from the neighboring villages who were not part of the oil exploration team. While ONGC employees referred to hydrocarbon operation areas as zones of interest and to the townships as the ONGC quarters, they called the world outside simply Assam. The CSR officer explained that Assam was a "difficult place to work" and talked about encountering a "communication problem." Such language allowed the organization to focus on exploration and to become familiarized with the landscape without engaging with the political reality of the place or the experiences of the people.

When I asked the CSR officer about the communication problems, he explained, "There is a huge problem of misrepresentation. Groups come to the office and often misrepresent themselves to get development projects and funds that are earmarked for certain operational sites." "Then what happens?" I asked. "Once the local people get the funds, there is no accountability or monitoring mechanism. If ONGC wants to monitor development activities, groups who have taken the benefits retaliate and refuse to cooperate. Therefore, there is no monitoring of projects and activities."

The ONGC activities included building roads and bridges to expand the area of its exploratory and extractive operations. Scholarships were provided for village schools and medical camps, but these were not philanthropic. "We only invest in places where we can get some benefit," the CSR officer said, implying that funds were earmarked for villages where oil exploration and production were taking place. But there were frequent conflicts over these issues. For instance, on July 13, 2015, the All Assam Students Union (AASU) called for a hundred-hour *bandh* (blockade) against the ONGC for failing to set up a hospital in Sibsagar and not generating employment for locals. They also demanded that the ONGC stop using Napukhuri and Puranipukhuri, important historical sites from the Ahom era, as warehouses for machinery. As the AASU stopped all ONGC operations in the area, security forces were called in to control the crowd.[2] In 2013, the police arrested over fifty residents in Nazira when they demonstrated outside the ONGC township to demand employment opportunities for local residents.[3] Similar encounters between security forces and protesters were not uncommon.

Although ONGC strictly prohibited employees from speaking out about political matters, they were unable to stop the consequences of the hydrocarbon exploration and the survey that took place in the towns and villages.

In 2008, the science and environmental magazine *Down to Earth* reported on people's experiences of ONGC seismic operations in Sibsagar district. These seismic surveys had damaged an eighteenth-century amphitheater from the Ahom era and other properties: "The survey at the Rupohipathar oil field caused several 10m wide cracks on the walls of the monument, called the Rang Ghar. . . . The residents living in the adjoining areas allege the oil giant is carrying out explosions far above the permissible 18m limit, leading to damages in residential buildings and other structures."[4]

Some years later, a journalist visiting the area wrote: "Some prime property around these tanks [historical sites] has been occupied by either the Indian Army or the Oil and Natural Gas Corporation Limited (ONGC India)—finding these lands empty when they first came here and with no security required to protect their equipment, lands were easily usurped. Incidentally, the ONGC was born in Sibsagar, and continues to operate in full swing. Its multibillion dollar market capitalisation has not moved the public sector company to bring the profits to its place of birth."[5]

Oil exploration takes place in the middle of paddy fields, in people's backyards, and adjacent to heritage sites like sacred temples and water bodies. Thus the notion of "communication problems" that the CSR officer stressed between ONGC employees and local residents seemed to simplify the complex ways hydrocarbon exploration deeply impacted the social world of the people in foothills of Assam. It was local knowledge that residents from the adjoining villages and towns were regularly frisked by the armed security forces, and the ONGC's construction of new culverts and bridges in a village meant that new oil wells were discovered in the vicinity. In addition, the high traffic of ONGC tankers and geologists traveling across the foothills meant that security forces were present in phenomenal ways. Given the volume of work, the constant demand for workers, contractors, and local partners from the foothills required identifying the *bhal manu* (good person) in the towns and villages. The foothills thus emerge as an important site of engaging with the concept of profiling and producing a carbon citizenship regime.

CARBON CITIZENSHIP

The presence of security forces, the history of militarization, and the heightened hydrocarbon operations distinctly configured the lives of people in the foothills of Assam and Nagaland. Locals constantly encounter or participate physically or imaginatively in the activities taking place around them. This means that the hydrocarbon operations in Assam and the coal mines in

Nagaland, as well as the tea plantation economy, produce the foothills as a rich resource hub, intimately shaping the lives of the people who live there. The ONGC's strict injunction to avoid politics, combined with its ongoing struggles with local workers and contractors, forced me to ask some basic questions. How did they seek out "good, reliable contractors" (a term the CSR staff used) from the adjoining villages and towns while avoiding the social space and values? On what basis did they qualify residents as trustworthy, responsible technicians? These queries allowed me to trace the relation between militarization and resource extraction in Northeast India and to highlight how hydrocarbon exploration in the foothills of Assam and Nagaland employed frameworks that were deeply militaristic, thereby creating a securitized carbon citizenship regime.

There are two ways to understand the production of carbon citizenship. In the first, foothill residents are required to produce documents, papers, identity cards, and affiliations with institutions, villages, or associations. This is often impossible, because many people have no documents. For example, women traders from the villages in Nagaland who go down to the weekly markets in Assam to sell produce do not have identity cards. As they cross the security checkpoints every week, they undergo a local process of profiling by security forces. This involves routine interrogation, frisking, and recognition of the traders' faces. At the discretion of the guards at the checkpoint, this test may depend on the women traders being respectful, subservient and fearful. The second test is mediated by power and violence and based on a violent form of improvised interrogation that is repetitive, exhaustive, and founded on uncertainty and humiliation. For example, even when people stopped at checkpoints are able to furnish identity cards, security forces harass and intimidate them nonetheless. One evening, I was stopped at a security checkpoint. I was barraged with a series of questions about my presence in the foothills. When the security forces checked my bags and found a packet of fresh bamboo shoot, they immediately assumed it was ivory. "So, you are smuggler!" they commented gleefully. Although their joy in catching an ivory smuggler was momentary, they further detained me. They began to ask me why I had cut my hair short. "You look like a like a boy," one of the officers commented. Pulling away my field informant to a safe distance, they began to harass him and interrogate him about my identity, suspecting me of being a member of a Naga insurgent group. This second test is designed to distinguish troublemakers from those who are trustworthy; as one might guess, it is extremely violent and terrifying. Both these tests are based on the intentions of the security forces to harass and the desire to scrutinize the practices of the residents on any given day. Both these tests are based on everyday assumptions, stereotypes, information, and rumors about the people and ethnic groups who live there.

These tests, founded on loyalty and trust, reinforce the division between outsiders and insiders. For Indian security forces posted here, this distinction is based on national territorial rootedness. Residents of the foothills are often put under greater pressure to prove they are good citizens despite living in a region that is unsafe and dangerous. In military logic, profiling reinforces the dominant national image of the good Indian citizen. Such securitized profiling crosses citizenship with a militaristic understanding of rootedness. In the foothills, people from other parts of India are regarded as detached from local history and social relations, and therefore not likely to create trouble. This is not the case for residents who share a larger matrix of kinship ties and networks, and who claim a history of belonging to the foothills. These bonds are perceived suspiciously. Throughout my fieldwork, I was categorized as a researcher belonging to the state of Nagaland and interrogated. Clearly, carbon citizenship tests are predetermined by ethnicity and political dispositions in the foothills.

My experiences of the foothill landscape were primarily through security checkpoints and the men in uniform posted there. Security forces stationed to protect the oil drilling sites and tea estates created a mechanism of profiling and categorizing people. It was routine to take addresses, names of villages, phone numbers, and vehicle registration numbers of travelers passing through the numerous checkpoints. These techniques of collecting information came from a long history of militarization and counterinsurgency; they are very similar to the categorization of groups as combatants and adversaries in the militarized history of Northeast India. The militarized profiling practice of the hydrocarbon exploration areas in the foothills fed into the extractive resource operations. The security profiling and ONGC classifications of the *bhal manu* reinforce each other's concepts of desirable people.

ANTI-NATIONAL

"I will tell you from the beginning," said Dr. Gogoi, a medical doctor from Sonari who started his practice in 1974. As an important coal town, Sonari was a hub for traders, insurgents, travelers, and contractors. When the government of India launched Operation Bajrang, a counterinsurgency program across Assam in 1991, thousands of civilians in Assam were profiled as suspicious persons, and many were executed. The Hindi term *bajrang* is associated with characteristics of masculinity and refers to "one having limbs (anga) [as hard as a] thunderbolt/diamond (vajra)" (Lutgendorf 2007).[6] It is a metaphor for strength, courage, and bravery. Dr. Gogoi was one of the victims of Operation Bajrang. His frequent travel to Naga villages in the hills and his association with poor patients in the foothills at the height of the counterinsurgency operations launched by the Indian security forces put him in the security category of the suspicious figure who was a sympathizer with the politics of dissent in the foothills.

He was suspected of sympathizing with the United Liberation Front of Assam (ULFA), an armed resistance group in Assam. Stories about ULFA and National Socialist Council of Nagaland (NSCN) are rife in the foothills. Even shopkeepers and bus drivers in the villages and towns seemed to recollect interactions with them. According to Dr. Gogoi, he met the "boys"—the insurgents—in the hills in the 1970s:

> When I was in a village in the mountains attending to an ailing woman, there were some men who were sitting and watching. But how could I know who was underground? There was no stamp on their forehead. I used to treat bullet injuries sometimes; they were local people and not undergrounds. The police came to me and told me to report such cases to them first. I told them, "Why should I do that? I am saving lives." [I thought,] "When I find suspected persons, I will inform the police." But many of these bullet injuries were routine accidents that took place during hunting trips.
>
> Around two years ago, there was an encounter between the two insurgent groups in the foothills. A rocket launcher fell on a pastor's compound and around seven to eight people were injured. They all came to my clinic, and I treated them. After they left, the Indian army came in civil dress and interrogated me. I said that I treated splint injuries, and the people were lucky because the bomb exploded on the roof so there were few casualties. The army wrote down everything. I told the army, "How could the injured go to the medical college in Dibrugarh? They had no money." They asked if any underground came for treatment. I said no, and showed them the number of patients and their names.
>
> In Longwa [a village in Nagaland], there is an Assam Rifles camp. [When I was there] they came and surrounded my hut. They accused me of coming to the village to give medicine to the underground. I told the security forces to ask the people why I had come there. I asked the security force, "Do you see

the situation? People have been saved because I have come here. I am running out of medicine and saline." I took them and showed them the patients who were on drips—there was a row of beds, and people were sleeping on the beds.

Eventually, Dr. Gogoi was detained by the Indian security forces. Dr. Gogoi's account of his night of torture captured the implications of being a suspicious person. As he talked about his treatment, he touched his head and face, running his fingers over the patches of burned and shriveled skin. He described the interrogation:

On December 28, 1990, they came for me. One of them was my youngest sister's friend in the university. I took my own car to the camp. It was very cold. My friend, another doctor, was also taken to the camp along with me. We were made to sit on the ground inside a tent. My friend was taken away, and soon after that I heard his screams. I knew what was coming.

After some time, an army major came inside the tent with a soldier. He was drinking. He asked me to take off my clothes. I was kept outside naked for thirty minutes. Then I was brought inside and electric wires were tied around my body. They gave me electric shocks. When I would scream, they would hit me with bamboo poles. I said, "I am feeling cold." The army major said, "You are feeling cold. All right," and he asked me to open my mouth. When I did, he poured chili powder in my mouth. I said, "Sir, I am dying; please give me water." The army major responded in Hindi, "Pani chahiye nah?" [You need water, do you?], and brought a jar of boiling water and poured it over my head. At that moment, I lost all sensation.

He started to interrogate me about my practice. I said, "I treat all kinds of patients." There were people who took me to the jungle to treat patients and I could not refuse. My job was to treat people. I told him, "You cannot ask me where I go because at that time, I was blindfolded and if I am told to go, then I have to go. I have a family to look after as well." Two days before I was taken to the army camp, on December 27, 1990, President Rule [martial law] was declared in Assam. All the big traders and rich contractors in the foothills gave the army our names. They said that the ULFA came to Sonari and sat with Dr. X [the doctor friend] and Dr. Rajan Gogoi. So, you can go and talk to them.

The army major said, "You are big *dada* [brother] of ULFA, we know. They call you *dada*." I replied that I was a doctor. I said that I lived near the hills, so I also traveled to the hills. I was bound to go to the hills. The army continued to interrogate me and tortured me for one and half hours. They brought my doctor friend into the tent. He was naked. They lit a gas stove and made him sit on the flames. He fainted. They poured water on him and tortured him again. The army major was very drunk by this time, so when it was my turn, he could not torture me too much. But my friend was in a terrible state.

Dr. Gogoi's story is symptomatic of a citizenship determined by ethnicity, political dispositions, and security profiling. For the Indian security forces, the bodies of the residents, or what the ONGC CSR officer called the "public," were a site for marking power, violence, and authority. The distinction between insurgent and civilian was framed through the eyes of the Indian security forces. For example, as part of the counterinsurgency operation, police checked the elbows of civilians at every checkpoint. They believed insurgents had rough elbows and other signs on their bodies as a result of rigorous training. Young men and women were forced to kneel as the security personnel checked each person for signs of being an insurgent.

Often, the lines blurred between a suspicious person and a well-trained body. Those who were physically fit and healthy were often taken to the security camps and brutally tortured in interrogation sessions. Dr. Gogoi recollected the violence he had witnessed during Operation Bajrang in the foothills: "Around these villages, the army tortured many athletes. Because they were fit and looked strong, the army destroyed their bodies. The football players' shins were smashed, the runners' feet were broken—people stopped playing sports. In a neighboring village, there were so many good football players—the army tortured all the boys. I questioned the army about their actions, and they said, 'There are orders from above. We are helpless. We do not know the people but there is information. . . . We act on the basis of this information.'"

"The Indian army has killed so many people here, it is uncountable. . . . They first torture and then interrogate. What is this?" Dr. Gogoi became agitated; the stories of the violence and torture, like entrails and organs bursting out from smashed bodies, left him mute with rage at times. The Indian security forces had tortured a young boy from Dr. Gogoi's neighborhood. They tied him upside down and shattered his thigh bones. Then they shoved bamboo sticks through his thighs. The young boy survived the torture but was crippled for life. The security forces, according to Dr. Gogoi, only arrested "the local people here. The others, who had come from outside the region and settled down here, were interrogated and released. They were not tortured or killed." One of the reasons the security forces targeted local residents was to extract information about insurgent groups like ULFA and NSCN. Family members, relatives, friends, kin groups, and neighbors—all categorized as local—were strategically arrested, interrogated, and tortured throughout the decade.

PROFILE WORLD

Dr. Gogoi recollected that those who came from other parts of India were only questioned. It appeared that no one from the trading communities was tortured or killed by the security forces. The distinction between the local residents who were tortured and the traders who were perceived as loyal citizens by the Indian security forces was stark. To show that the Nepali people also belonged to the foothills, Dr. Gogoi mentioned that many Nepali residents were also tortured because "information" was passed to the security forces. "Who did that?" I inquired. Dr. Gogoi speculated that it was the Marwari and Bihari traders: "See, they [Marwari and Bihari traders] are also involved with the underground and they also work with them. [But] the army captured the Assamese people here—especially the Ahoms, Kachari, and others because for the army, ULFA means Assamese. There is no difference between the Assamese and the ULFA."

According to Dr. Gogoi, Indian security forces spared the trading and business communities from other parts of India because they were not perceived as suspicious. For example, the Marwari community in Assam and Nagaland held large shares in the tea plantation economy, owned major businesses including gas stations, timber companies, and construction companies. They also had a monopoly over all the major commodities and other supplies that arrived in Northeast India from different parts of India.[7] Yet the security forces maintained that commerce and conflict were distinct paths that did not intersect, contrary to the accounts of local people in the foothills, who highlighted that traders benefited immensely from selling goods to the insurgent groups in the foothills.

One evening, I met an elderly Marwari shop owner in Sonari town. He narrated how Naga "undergrounds" regularly came to his shop at the height of the Naga insurgency in the 1960s and 1970s to buy olive-green cloth for uniforms. Then he described his timber trade in Nagaland during the 1960s. There was an elephant camp outside Sonari, and he regularly hired the elephants to go up to Nagaland and carry the timber from the forest. In the middle of our conversation, his son came in and took away his father away, leaving the story of the timber and elephants incomplete.

The Marwari traders remained distant and turned down my request for meetings. "Moi khan iku najane" (We are not aware of anything), they told me and fell silent. Yet their social networks and scale of operations said something else. Trading communities like the Marwari were among the biggest economic beneficiaries of the conflict economy in the foothills. All the major shops and supply chain networks were controlled by them. Many Marwari families traced their family business to the early part of the twentieth century, and like other ethnic groups in the foothills, they encountered both state and nonstate actors. They belonged in the foothills and called the place their home. Yet, in the eyes of the security forces they were reliable citizens.

While residents like Dr. Gogoi were perceived as suspicious and disloyal, his neighbors—the Marwari—were appreciated as good citizens. These profiling practices revealed a militarized reality. The model of profiling carbon citizenship erased the complexities of social relations in conflict economies. In reality, a major part of the funding for coal mining in the Naga village came from Marwari business groups settled in Tinsukia, Jorhat, Guwahati, and Dimapur. Dr. Gogoi's town was not far from the Naga coal mining villages. Approximately eighteen kilometers away, the movement of traders and goods indicated that everyone, including Ahom, Marwari, and Assamese business groups, was seeking new opportunities and territories to expand the coal mining areas.

Every trading group, whether Marwari or Ahom or Naga, sought to secure important government contracts and collaborated with local contractors to supply construction machines, vehicles, and other goods to the extractive industries (oil, coal, and tea). To manage their trading activities, they maintained working relations with various political actors. Therefore, conversations about Marwari communities, who, in Gogoi's view, enjoyed the protection of the Indian security forces, could be applied to the trading communities who served the interests of the extractive resource regime and reinforced the security regime of the carbon citizenship test.

CONCLUSION

The interaction between the Indian security forces who protect the oil exploration sites and townships and the residents outside these oil worlds largely consists of force and threats. The zone of interest erases people and social relations, including everyday experiences of militarization, and instead employs technical language in overseeing oil exploration. This produces the foothills as a resource-rich hub free of conflict or obstruction to the goal of extracting those resources. Indian security forces have used surveillance to produce a militarized profiling system of carbon citizenship. A process by which model citizens are shaped using definitions of the *bhal manu* as laid out by the hydrocarbon practices and illustrated by the ONGC CSR office shows a system that erases the violence and militarized history of the place.

Against the backdrop of a disturbing political history of violence, the everyday tasks of governance are left to the Indian security forces, whose sensibilities are militaristic. The insidious logic of the Armed Forces Special Powers Act (1958) and the Disturbed Areas Act (1955) is the reproduction of the "disturbed area" as a geographical location (Northeast India) and a social category (northeastern people). Circular logic predicts that places inhabited by suspicious people will eventually become a disturbed area, and those inhabiting the disturbed area will naturally become suspicious people (Kikon 2009: 272). Suspicion occupies a space between law and uncertainty here. Judicial and policing systems throughout the modern state system are organized on the presupposition of suspicion and uncertainty (Asad 2003). In situations where the state categorizes an entire geographical region as a "disturbed area," the lens of suspicion scans every single inhabitant. The legal regimes and systems of policing that check the elbows of youth, smash the ankles of football players, torture doctors, and kill residents blur the boundaries between protecting sites of resource extraction, profiling good citizens, and securing boundaries around oil rigs. This militarized framework of profiling people and protecting resource extraction produces carbon citizens in the foothills of Northeast India.

Past, Present, Future

WHEN I started fieldwork in the foothill border of Assam and Nagaland, I came across a school in a Naga village that had been built on the ruins of the ONGC's oil exploration offices after the work was abandoned in 1994. The following vignette from my field diary captures the lives of people who reside in this carbon landscape and the connections between the harsh reality and their aspirations, irrespective of how violent or reprehensible they may appear.

The English School was a prominent landmark. When I visited, there were six teachers and one hundred students who came from the surrounding coal mining villages, tea plantations, and oil rig sites. They came from diverse ethnic backgrounds: Naga, Bihari, Nepali, Ahom, Mishing, Adivasi, and Bengali. The principal told me how they had used the foundations from the old ONGC building to erect a school for the village and had added classrooms as the number of students increased over the years. They also converted part of the old ONGC offices into teachers' quarters. The concrete road that once connected the offices to the oil wells was still in good shape and functioned as the school playground. On weekends, the school watchman's family and the teachers who lived in the school compound dried paddy, grains, firewood, and clothes there.

There was an Assamese school in the vicinity, but parents preferred to send their children here because instruction was in English. Teachers told me that families wanted their children to learn English. In this village surrounded by hydrocarbon sites and tea plantations, the parents were aware that the higher management of the oil and tea companies carried out their duties in English. They wanted their children to learn to read and write English, but the principal explained that there were challenges to teaching English. She reflected, "It is not an easy job to teach [them]," because

Schoolchildren walking home (Assam and Nagaland border)

"several students come from families where parents are daily wage labor-ers," thus implying that the parents were illiterate and could not help with homework.

To make sure that students spoke in English when they were at school, the teachers used two strategies. The first was "wearing the sign." When teachers caught students speaking a language other than English, they inflicted "double punishment." This meant that both parties—the speaker/s and the listener/s—were made to wear cardboard signs reading, "I am a donkey, I speak in Nagamese" or "I am a monkey, I speak in Assamese/ Bengali." Each signboard had one or more languages scribbled next to the "I am a monkey/donkey" sentence. The only way to remove the sign was for the teachers to pass it to other students caught speaking a language other than English.

The second method—also a punishment—was known as "showing the injection." The principal kept a disposable hypodermic needle in her office for this purpose. One day, I saw a teacher apply this method. She brought the needle close to a student's arm and simulated a pre-injection procedure. She lifted the child's sleeve and then looked at the student, a small boy not more than five years old, and told him not to cry. She explained that it was

Punishment for speaking Nagamese and Lotha (Nagaland)

necessary to give him the shot to make sure he spoke in English. The children were scared of the injection, and the teachers purposely used it to frighten them. As the teacher brought the needle close to his skin, the child froze. As tears streamed down his chapped cheeks, the trembling child pleaded in Nagamese, "Nai nai" (no, no). The teacher pretended not to understand. Finally, the child managed to say the words "English, English!" and nodded his tiny head vigorously. The teacher allowed him to return to his seat.

It was a very unpleasant sight. Disturbed by the harshness of the punishment, I asked why such harsh methods were necessary to teach English. The teachers said they had to "push" the students to learn English. In many ways, the English School, built on the ruins of the ONGC office, was a microcosm of the foothills. Fantasies and aspirations were fueled by violent and traumatic experiences with state agencies and resource extraction operations. In the midst of a carbon future in Nagaland and demands for equitable carbon royalties for the people in Assam, many residents in the foothills who barely managed to make ends meet began to believe in the prospects for a secure hydrocarbon future. The expanding ONGC surveys for oil and gas in Assam and the increasing number of coal mines in Nagaland worked a spell on the residents, making them victims of a grand fantasy where the foothills would be an everlasting hub of hydrocarbon—a place where oil, natural gas, and coal will bubble away on the surface of the soil forever, and the people will smile, speak in English, and soak up the profits.

In reality, much like the teachers at the school who prepared the children to become better students and speak in English, state agencies like the ONGC and the security forces swarmed to the foothills and installed oil rigs and checkpoints to train residents to become better citizens, rendering loyalty and service to the nation. As these sites became resource hubs of energy and carbon, they generated dreams amid misery on the ground: Mr. Kithan, who lived in a village without electricity yet fantasized that he was guarding precious oil beneath his field; Yan's husband, who lived in an oil dream world; Alem, who negotiated the patriarchal world of a Naga family and managed the coal mines; and the small child who cried out, "English, English!" to show he could speak the language.

The daily surveillance and conflict I described in the introduction highlighted the lives of people in the foothills. The landscape as a hub of hydrocarbon shaped the political lives of the ethnic groups who lived there. Hostilities increased during coal mining seasons and agricultural periods of sowing and harvesting. These experiences provide important insights into how land and resources are perceived, managed, and controlled in this region, including the way they legitimize the power to validate ethnic histories and notions of belonging, culture, and memories. By this logic, these entitlements justify seasonal friendships, adoptions, and alliances that are established to carry out extractive operations such as coal mining, sand mining, and new plantations. This explains why, despite the absence of electricity, schools, primary health care, or water in numerous villages, people are determined to stay here. In these villages, they organized themselves along respective ethnic groups, clan, or tribe lines in matters relating to land deals and resource extraction activities.

The coal mining activities in the Naga villages and mountains were privately financed and outside the ambit of the ONGC and state authorities in Assam. Unlike the ONGC geologists who drilled for oil and gas in Assam and enjoyed the support of the state, the coal mining villages sought protection from various powerful bodies, such as tribal councils, student unions, rich politicians, and nonstate armed groups. Furthermore, unlike oil operations in Assam, which sought to project their activities as part of the national interests and vision of meeting India's energy demands, coal trading alliances invoked a cultural and historical past of camaraderie and kinship. Local legends, such as the one about the two brothers described in the introduction, along with social relations, exemplify ways of navigating anxieties and tensions.

Though ONGC experts saw oil and gas operations in Assam as technical activities detached from social and political realities, in the foothills the spatial worlds of oil exploration and the social lives of the people in towns and villages collided, often in violent ways. In Nagaland, coal was used to assert territorial power and ethnic alliances. Nevertheless, both oil and coal activities framed the relationship linking extractive operations, violence, and militarization in Northeast India. The traffic and management of these resources illustrated the role of multiple authorities, including state and nonstate actors. These complex militarized relationships became visible through social, spatial, and political accounts of living with oil and coal in the foothills.

NOTES

FOREWORD

1   *Zomia* is used to signify disconnection and intractability in the face of
    expanding state power across the hilly regions of mainland Southeast
    Asia adjoining areas of northeastern India that Kikon is writing about.
    See James C. Scott, *The Art of Not Being Governed: An Anarchist History
    of Upland Southeast Asia* (New Haven, CT: Yale University Press, 2010).

INTRODUCTION

1   Borders, as Anna Tsing (1993: 72) notes, are about defying the rules and
    the demands of state-administered rules. Taking into account Tsing's
    understanding of the border, I interrogate who wields power in such
    spaces. From physical geographers who initially guided my perception
    of the foothills (Prescott 1987: 13; Newman 2006: 144), and later schol-
    ars who wrote about borderlands (Van Schendel 2002; Nugent 2002;
    Zamindar 2007), I was able to understand the foothills as a space of
    extraction where the boundaries were at times physical, visible lines
    and at other times social spaces for establishing networks, social rela-
    tions, and alliances.
2   Anna Tsing (2003) describes how frontiers are produced as projects
    that create and offer geographical and temporal experience. The frontier
    in Northeast India, in a similar fashion, was produced to expand the
    military and economic projects of the British Empire in the nineteenth
    century. This meant that two kinds of wilderness—people and the
    landscape—had to be disciplined.
3   Arun Agrawal's (2005) work on the Kumaon hills in India describes a
    structured community where power, relations, and politics are produced

and reproduced through a disciplinary environmental practice he calls "environmentality." Agrawal describes three governmentalized sites that emerge from the relationship between government and locality: localities, communities, and subjects. He notes how governmentalized localities create relationships among centers, localities, and subjects woven together by state power. Unlike the Kumaon hills, the foothills of Northeast India are founded on a violent history of armed conflict and militarization that dictates and shapes everyday experiences of extractive practices and the networks of power. For this reason, I focus on the shifting networks of power and sovereign actors (state and nonstate actors), describing how people along the foothills live out their lives in such precarious places.

4   Avik Chakraborty, "AASU Protests OIL, IOC Merger Move: Activists Submit Memorandum to Sonowal," *The Telegraph*, August 5, 2017.

5   The United Liberation Front of Assam (ULFA) describes itself as a revolutionary organization and believes in the right to self-determination. ULFA insurgents have waged an armed struggle against the Indian state with an aim to liberate Assam and establish a sovereign and independent Assam. Currently, the ULFA and the government of India have signed a cease-fire agreement and are currently engaged in a political negotiation. "ULFA as a Revolutionary Organization," ULFA website. The archived website is available at www.oocities.org/capitolhill/congress/7434/ulfa .htm (accessed March 21, 2018).

6   This insight is from personal communications with Sarat Phukan in Guwahati in 2017.

7   Interview with Sarat Phukan conducted January 15, 2018, in Guwahati (Assam).

8   The famous slogan "We will give blood but not oil" (Tez dim, tel nidiu) was coined in the early 1980s in Assam as a distinctive political assertion of the right to control natural resources and the intent to agitate against "outsiders" taking resources and land from the people.

9   Sangeeta Barooah Pisharoty, "Oil and 'Outsiders': Outrage in Assam over the BJP's Decision to Privatise Oil Fields," *The Wire*, July 10, 2016, https:// thewire.in/politics/bjp-assam-oil-privatisation.

10   While the term Maan-Singhpo appears as a name of a person in the story, according to historical sources it was an alliance between the Singpho tribes and the Burmese kingdom. Maan is the name for the Burmese people in Tai Ahom language. Hill groups were often conscripted as soldiers by neighboring kingdoms during wars.

11   The Ahom kingdom comprised the plains of eastern Assam for six hundred years. During the reign of the Ahom kings, social and political networks were established with the surrounding hill people through marriage alliances, trade, and tributes. Across the foothills, stories of

solidarity between the Ahom kingdom and the Naga people in the hills were frequently invoked to indicate goodwill and resolve conflicts.

12 The South Asia Terrorism Portal (SATP) is maintained by the Institute for Conflict Management in New Delhi. Two of their core areas of research include continuous appraisal of internal security and the state's responses in all areas of existing or emerging conflicts in South Asia; and planning for development and security in India's Northeast. The research center views any assertion of rights to self-determination and the resultant insurgency as terrorism. For a detailed list of the armed groups and their locations, refer to www.satp.org (accessed February 14, 2018).

13 Directorate General of Hydrocarbons, *Basin Information: Assam-Arakan Basin* (Noida: Ministry of Petroleum and Natural Gas, Government of India, n.d.).

14 Amnesty International, "Public Statement—India: Concern over Reported Unlawful Killings by Security Personnel in Assam," January 26, 2007.

15 Sanjay Barbora, "'Milakpani te ahibo, sopna te dekhibo,'" *Himal South Asian,* January 23, 2007.

CHAPTER 1: STORYTELLERS

1 Gazettes are remnants of colonial administrative notifications that are published by the state and central governments in India. The gazettes contain announcements of departmental projects and official notices from the government.

2 *Adivasi* literally means indigenous people who are original inhabitants of the place (in Hindi and Bengali). But in Assam and other parts of Northeast India, this term specifically refers to the indentured tea plantation workers brought from central and eastern India in the nineteenth century by colonial administrators.

3 It is important to remember that since the 1990s, due to the increasing pressure to reduce labor costs in the tea plantations, management have favored casual labor. Casual workers often do not have access to benefits such as plantation medical centers, housing facilities, or the grain distribution system. So, they seek employment as daily wage laborers across the foothills in the coal mines, paddy fields, and rubber plantations.

4 Behal (1985: 19–20) notes that as part of a "disciplining" regime, workers were compelled to reside on the plantations, and their mobility outside the plantation was severely restricted. The employer-employee relationship on the plantation, which made them completely dependent on their respective employers for their lives, was paternalistic, and the workers were expected to be submissive.

5 When I inquired about the festivals his family celebrated, Christmas was first on the list. It was the most important festival, followed by New Year

and Easter Sunday. "But the Adivasi families join the Assamese people during Bihu [the harvest festival in Assam]. They visit one another and dance together during festivals." He smiled. His sons, who were sitting with us, shared how they visited their Assamese friends during Bihu to beat the drums and participate in the festivities.

6   Today *coolie* is a derogatory term used to refer to a particular disposition attributed to plantation life. The origin of *coolie*, according to historian Jayeeta Sharma (2009: 1304–10), lies in the Tamil word *kuli*, meaning "wages." The coolie labor force was created as replacement labor after the British Empire's emancipation of slaves in 1830. Most of the coolie labor force came from central and eastern India. They became part of the labor force that was sent to the tea plantations in Assam and sugar plantations in the Indian Ocean to replace the African slaves. Also see Kaushik Ghosh's work (1999) on the indentured labor market in colonial India and the process of physically and politically displacing the population.

7   There is a rich debate about who is indigenous, aboriginal, and tribal in India, but my main focus is on the status of the Adivasi people in Assam. For the vibrant academic debates on the topic of tribes versus indigeneity in India, see Xaxa 1999; Beteille 1986; Ghurye 1963; Roy-Burman 1983.

8   Liisa Malkki (1992: 26) notes, "One country cannot be at the same time another country," as she describes the emergence of national boundaries and spatial sovereigns in the twentieth century.

CHAPTER 2: DIFFICULT LOVES

1   See also Dolly Kikon, "Operation Hornbill Festival 2004," Seminar #550, www.india-seminar.com/2005/550/550%20dolly%20kikon.htm.

2   According to Sara Ahmed (2003), groups who apply the language of love do so to invoke passion and defend the nation against others. She argues that the bond among members of such groups relies on the transference of love to the leader. In this process of shared orientation, the ego of the leader is pushed toward the loved objects: the nation and homeland. But more importantly, the process also assumes the characteristics of a lost or threatened love object. Therefore, the loss of or threat to the homeland conjures a deep sense of mourning and grief. Sorrow becomes an important expression of love.

3   There are numerous accounts of Naga men and women who have died in the Indo-Naga armed conflict for the love of the Naga nation (Kikon 2004; Banerjee 2008; Sanyu 1996; Iralu 2001). There are memoirs of Naga soldiers, families, clans, and at times entire villages that were devoted to the Naga people's struggle for independence. Thus—to borrow a phrase from Sharika Thiranagama narrating the Tamil struggles in Sri Lanka, "One must either love it or feel obligated to love it" (2011: 19)—the notion

of love, applied to the struggle for a Tamil homeland in Sri Lanka, emerged from social pressure to reinforce a particular set of values and beliefs. This act of devotion to the homeland coexisted with a threat of transgression that came from members within the community. In such situations, love for a homeland is perceived as a transgression when there are, in Mary Douglas's terms, "lapses from righteousness." Douglas explains that the act of transgression operates in a logic seeming to suggest that "which is not with it, part of it and subject to its laws, is potentially against it" (1966: 3–4). The process of identifying the act of purity and punishment is set in place to impose order.

4   Demands for exclusive homelands in the hills and plains often led to ethnic conflicts across several federal units in this frontier region (Baruah 2005). To a large extent, states are responsible for the forms in which politics of recognition and claims to territorial homelands emerge and are articulated. Povinelli (2002: 176) refers to the distinction between "us" (the state and its perfect citizens) and "them" (the aboriginals, indigenous groups, tribes, and all the other categories) as "the politics of repugnancy." Yet increasing contestations over land rights and claims for recognizing spatial identity emerge from contestations over colonial boundaries, postcolonial territorial modifications, regional border disputes, and multiple claims to homelands. These demands are closely linked to the politics of governance and the way people understand and negotiate with the state. Adopting spatial practices and demarcations as naturalized categories reinforces state practices and the spatial order, which reproduce forms of inequality and oppression (Spivak 1999; van Schendel 2002; Cohn 2004; Ferguson 2006).

5   Pratiksha Baxi highlights the connection between the family and the state. She notes how the family and the state have consistently worked together to establish the limits of intimacy and thus reinforce the institution of the family, and she notes that everyday narratives of love are intimately "entangled in procedural law" (2009: 1). Therefore, it is important to recognize how notions of sovereignty and limitation are defined in relation to love.

6   Unlike the camp refugees, the town refugees were not as concerned with purity, and they negotiated multiple identities and practices grounded in the social context of the township they inhabited. The town refugees embraced a creolized, hybrid, cosmopolitan identity and rejected identities such as refugee or Hutu and other similar ethnic or national markers. The manner in which the camp refugees perceived their town counterparts as rootless and impure illustrates how notions of belonging and spatial categories are intimately shaped by life experiences. The lives of the foothill residents are somewhat similar to the lives of the town refugees that Malkki (1995) describes. Yet there is no spatial distinction

similar to the refugee camp and the town for the foothill residents. Their homeland is where they live. It includes their villages, the coal mines, the tea plantations, the rivers and streams, or the oil exploration sites where different ethnic groups struggle to create a sense of belonging every day.

7   Culture has been an important locus for naturalizing power, practices, and hierarchies. According to feminist anthropologists Sylvia Yanagisako and Carol Delaney, "Culture is what makes the boundaries of domains seem natural. . . . [This perception of culture is] what gives ideologies power, and what makes hegemonies appear seamless. . . . [T]his creative dialectic of the concept [of culture] depends on our commitment to use it as an incitement to continually rethink what is same and what is different, how they are so and what this means" (1995: 19). Thus, the field of culture has become a productive site for discussions of nature, power, and culture.

8   "Sati Jaymati," Assam Info, www.assaminfo.com/famous-people/51/sati -jaymati.htm (accessed February 7, 2016).

9   Anthropologist Alban von Stockhausen (2015) shows how this legend continues to play an important role in negotiating local identities among the Konyak Nagas. Highlighting contemporary rivalry among Konyak clans to claim her allows us to understand the intersections of oral narratives, colonialism, and the written script in relation to identity politics and oral traditions.

10 Jennifer Cole (Cole and Thomas 2009: 110) explains the notion of the Christian representations of self-sacrificing love.

11 Here *dhorom* refers to the Christian faith. People use the term *dhorom* to talk about Christianity in Nagamese.

CHAPTER 3: STATE LOVES

1   As he explained the problem, it became apparent that the Nagaland government had actually distributed some government subsidies for wells, but internal village politics prevented the village from obtaining clean drinking water: "There were two houses in this village with wells, which served as the drinking source for many households in the village. The hill government wanted to give a subsidy for a well. Instead of giving it to these two households, it distributed it to some other houses where the well was not fit for drinking. There were four well subsidies for this village. The two wells that got the subsidy are not fit for drinking, while the other two wells have not been dug as yet. See, the commonsense decision would be to give the subsidies to those houses that had wells and were supplying drinking water to the village. But as we all know, in the Nagaland government, it's the party that matters and all other things don't matter. Even if there is a valid reason to support a cause, the party has first priority and help is given to the party members."

2   An Assamese journalist who later became press secretary for the chief minister of Assam coined the term *Seven Sisters* after the formation of Meghalaya in the 1970s. It encapsulates the wishes and aspirations of a local elite—especially those in Assam—to showcase the unity of the indigenous communities, despite the formation of ethnic states (Deka 2008). The Seven Sisters are Assam, Nagaland, Manipur, Tripura, Arunachal Pradesh, Meghalaya, and Mizoram. However, after the inclusion of Sikkim as part of Northeast India, the eight states are generally referred to as member states of the North Eastern Council, the agency overseeing the economic and development programs in the region.

3   This is an important consideration with political ramifications. Most hill states are primarily home to one ethnic group. Though it is home to many tribes, Nagaland is seen as belonging to the Nagas; Meghalaya to the Jaintia, Khasi, and Garo tribes; Mizoram to the Mizo; and Arunachal Pradesh to the Nyishis, Adi, Apatani, Galo, Mishmi, Sherdukpens, and others.

4   This is most evident in the contests around political representation and control over resources in the autonomous councils meant for certain ethnic groups. Under the Sixth Schedule of the Indian Constitution, the government of Assam, in consultation with the central government, can allow for a devolution of powers to certain tribes that live within a contiguous area. Under the aegis of the Sixth Schedule, powers of land transfer and ownership are often passed on to the autonomous councils, where nontribal groups have little political representation. The 2012 violence in western Assam's Bodo Territorial Council and 2013 violence in areas under the Rabha Territorial Council are prime examples of this fractious politics.

5   Sanjib Baruah, "Northwest by Northeast: A Tale of Two Frontiers," *Open-India*, July 17, 2009.

6   Sanjib Baruah, "Stateless in Assam," *Indian Express*, January 18, 2018.

7   Northeast Tourism, "Nagaland: The Land of Festivals," North Eastern Council, http://northeasttourism.gov.in/nagaland.html#sthash.prSIMxwP.dpbs (accessed March 2018).

8   Department of Tourism, "About Nagaland," Government of Nagaland, http://tourismnagaland.com/main/?page_id=60 (accessed May 2016).

9   For details on the constitutional amendments including their goals and motivations, see the Statement of Objects and Reasons, "The Constitution (Thirteenth Amendment) Act, 1962," National Portal of India, https://www.india.gov.in/my-government/constitution-india/amendments/constitution-india-thirteenth-amendment-act-1962.

10   For the current discussion, points 7 and 8 in the Sixteen-Point Agreement are important because they specify that no acts of the Indian Parliament may alter existing religious, social, and customary practices of

the Naga people in Nagaland: "7. Acts of Parliament: No Act or law passed by the Union Parliament affecting the following provisions shall have legal force in the Nagaland unless specially applied to it by a majority vote of the Nagaland legislative Assembly: (a) The Religious or Social Practices of the Nagas. (b) The Customary Laws and Procedure. (c) Civil and Criminal Justice so far as these concern decision according to the Naga Customary Law. The existing law relating to administration of civil and criminal justice as provided in the Rules for the Administration of Justice and Police in the Naga Hills District shall continue to be in force. (d) The ownership and transfer of law and its resources. 8. Local Self-Government: Each tribe shall have the following units of the rule making and administrative local bodies to deal with matters concerning the respective tribes and areas: (a) The Village Council; (b) The Range Council; and (c) The Tribal Council." Available at https://peacemaker.un.org /sites/peacemaker.un.org/files/IN_600726_The%20sixteen%20point%20 Agreement_0.pdf.

11   There is an extensive literature on the Naga armed conflict (Asoso 1974; Sanyu 1996; Lotha 2009; Baruah 2005; Iralu 2001). However, my project focuses not on the armed conflict in Nagaland but on the everyday lives of the residents who live on the foothill border.

12   While the Naga armed conflict started with the NNC, the Naga armed group was divided into two factions in 1980 with the emergence of the NSCN. The NSCN further split into multiple factions since 1997. All these groups assert that they represent the Naga people and are fighting for a sovereign Naga nation. As a gesture of support for the cease-fire agreement, Naga cultural and political associations in Nagaland and Manipur boycotted the Indian parliamentary election of 1998, arguing that the cease-fire agreement period should have focused on finding a political resolution and establishing a new Naga constitution rather than reinforcing the power and authority of the Indian state by pushing for an election.

13   The competing Naga insurgent groups' proposed Naga homeland covers a vast geographical area that stretches from the northwestern province of Myanmar to the plains of the Brahmaputra Valley in Assam. One can argue that the creation of Nagaland in 1963 with exclusive rights and guarantees for the Naga people was, in some small part, a concession from the Indian state.

14   Constitutional provisions, such as Article 371A, that guarantee special protections and rights regarding land ownership profoundly shaped the politics of tribal affinity and power in Nagaland. In addition, tribal institutions such as village councils, customary law, and other traditional practices are all involved in governing the hill states, including legitimization of extractive resource regimes like coal mines, plantations, and the timber business. Therefore, traditional structures and cultural

practices significantly contribute to contemporary state-building projects in Northeast India.

15 When respondents in the foothills referred to states, they usually meant their parent states of Assam or Nagaland. When they referred to the nation-state through which they claim formal citizenship (and violence), they meant India and, more specifically, New Delhi, since the capital city was seen to be the source of power and decision-making.

16 Mazumdar Prasanta, "Assam, Nagaland Lock Horns Again over Border Dispute," *New Indian Express*, May 6, 2015.

17 Public speech delivered by Mr. Rio, chief minister of Nagaland, in Chi village (Mon district) in 2010. I attended this meeting at the village during my fieldwork.

18 "Rio Calls for Bridging 'Constitutional Divide,'" *Morung Express*, May 4, 2007.

19 See the National Rural Employment Guarantee Act of 2005, http://nrega .nic.in/netnrega/home.aspx.

20 Sanjib Baruah, in "Nationalizing Space: Cosmetic Federalism and the Politics of Development in Northeast India," shows how these frontier regions are global hotspots of biodiversity and rich mountain ecosystems. As a result, focusing on road construction and increasing the number of roads and similar infrastructure might not fit a model of development. Pointing to the fragile mountain ecosystems and climatic conditions that cause landslides, Baruah (2003: 916) critiques ongoing developmental projects of the Indian state in frontier states like Arunachal Pradesh, referring to the Indian state's "goal of nationalizing the frontier space." For the logic of development projects in frontier areas like Northeast India, also see David Ludden's "India's Development Regime" (1992), where he shows how colonialism and the logic of the empire have intimately shaped India's perceptions about economic development. In the foothills, seasonal workers squatted on the damaged road with iron brushes and cleared dust and pebbles from the potholes and cracks, poured in hot tar, and patched the cracked roads. But the long monsoon and frequent landslides washed away the temporarily patched bridges and roads. These structures had been built using substandard construction materials, as contractors and officials pocketed a large share of the allotted funds. Given the sporadic road construction and limited life span of the roads, they become easy examples to signify the indifference of the state.

CHAPTER 4: THE *HAATS*

1 Prasanta Mazumdar, "Oil Wells That Brought Two Nagaland Villages Health Hazards," *DNA India*, September 18, 2011.

1   Public speech delivered by Mr. Rio, chief minister of Nagaland, Chi vil-
    lage (Mon district) in 2010. I attended this meeting at the village during
    my fieldwork.

2   In 1993, the resolution was later passed as a state law, the Nagaland (Own-
    ership and Transfer of Land and its Resources) Act of 1990, by the Naga-
    land state legislature (available on the state government website, https://
    www.nagaland.gov.in).

3   Although the attacks had receded in the last two decades due to cease-
    fire agreements with insurgent groups in Assam and Nagaland, the Indian
    security forces intensified military operations and surveillance during
    these national events as a precautionary measure. It was routine for
    administrators to seal off the foothill borders to prevent "antisocial"
    activities during these periods.

4   *Adivasi* means "original inhabitants of the land." Three terms—*tribal,
    Adivasi,* and *indigenous*—have definite genealogies in India. Sometimes
    these three terms are used interchangeably to refer to a single group.
    Yet they are highly contentious terms that continue to invoke debates
    about the politics of indigeneity and tribe in India. Here, it will be
    sufficient to note that the Adivasi in Assam are requesting scheduled
    tribe status under the constitution in order to access affirmative action
    programs.

5   Michael Taussig (1980: 117) underlines how stories about magic and
    sorcery are intimately related to sociality. Society members who face
    competition, envy, obligation, and other conflicts have often employed
    magic to address their concerns. Thus, for Taussig, sorcery is the one
    manifestation of "moral codes in action."

CHAPTER 6: CARBON FANTASIES AND ASPIRATIONS

1   The mining was poisoning streams and ponds. The scale of deforestation
    and the determination of Naga householders contradicted accounts of
    indigenous people as activists, resisters, or defenders of the resources.
    The resource extraction activities there highlighted how power relations
    and competing sovereign powers (armed groups, politicians, and indige-
    nous political actors in Nagaland) engineered the flow of resources.

2   The year 1997 was a watershed in Naga history. It signaled a cessation of
    hostilities between Naga armed groups and the government of India but
    also led to increased conflict between Naga armed factions, each seeking
    legitimacy and control of people and resources. Elected representatives,
    tribal institutions, insurgent groups, and government departments vied
    for the right to represent the Naga people.

3   Prasanta Mazumdar, "Oil Wells That Brought Two Nagaland Villages Health Hazards," *DNA India*, September 18, 2011; Nitin A. Gokhale, "Double-Barrelled Trouble," *Outlook*, October 12, 1998.

4   Minorities at Risk Project, "Chronology for Nagas in India," 2004, www .refworld.org/docid/469f38981e.html.

5   A Naga politician, M. Kikon, filed suit against the ONGC and other organs of the state, charging negligence, health risks, and environmental degradation resulting from continuing oil spills from the abandoned sites. M. Kikon sought a financial package as compensation for ONGC activities in Nagaland. Increasingly, natural resources, especially oil in Nagaland, became an important ground for mobilizing social groups, as well as an important agenda for political entrepreneurs. For details, see "PIL Filed against ONGC in Guwahati HC," *Morung Express*, August 28, 2011.

6   The Nagaland Coal Policy Act of 2006 did not regulate the number of coal prospecting licenses, coal mining leases, or small pocket deposit licenses. In 2008–9, the government of Nagaland came down hard on the villages for their coal mining activities.

7   Department of Geology and Mining, *Nagaland Coal Policy (First Amendment) 2014* (Kohima: Government of Nagaland, 2014).

CHAPTER 7: CARBON CITIZENSHIP

1   Upper Assam is very important for understanding the political and social imagination of contemporary Assam. Historian Yasmin Saikia's (2004) work on the Tai-Ahoms in Upper Assam highlights the importance of understanding how the landscape of Ujoni Assam (Upper Assam) shapes the people's longing for a lost homeland. She highlights how Ujoni Assam refers not to a geographical location but to an internalized intimate landscape in the memory of the people. Saikia's observation can be extended to the foothills and hills surrounding Ujoni Assam, where political fantasies also capture the foothill geography.

2   Prabin Kalita, "ONGC Operation in Assam Suspended after AASU Bandh," *Times of India*, July 13, 2015.

3   S. K. Hasan, "Protest over Recruitment at ONGC Nazira: AASU, AJYCP arrested," *Assam Times*, April 28, 2013.

4   Sudipta Nayan Goswami, "ONGC's Seismic Surveys in Assam Affect Heritage Monuments," *Down to Earth*, June 1–15, 2008.

5   Aheli Moitra, "Mung Dung Sung Kham," *Morung Express*, October 25, 2015.

6   The name Bajrangbali is associated with the Hindu god Hanuman. The term is associated with power, strength, and protection.

7   Historian Jayeeta Sharma (2011: 236) notes that the Marwaris are a well-connected, powerful business community in Northeast India. Originally from Rajputana in North India, they came to Assam in phases, but their identity was reformulated with the British colonial economy in Assam in the nineteenth century.

BIBLIOGRAPHY

Agrawal, Arun. 2005. *Environmentality: Technologies of Government and the Making of Subjects.* Durham, NC: Duke University Press.

Ahmed, Sara. 2003. "In the Name of Love." *Borderlands E-journal* 2 (3). www .borderlands.net.au/vol2no3_2003/ahmed_love.htm.

Appadurai, Arjun. 1986. *The Social Life of Things: Commodities in Cultural Perspective.* Cambridge: Cambridge University Press.

Arendt, Hannah. 1973. *The Origins of Totalitarianism.* New York: Harcourt Brace.

Asad, Talal. 2003. *Formations of the Secular: Christianity, Islam, and Modernity.* Palo Alto, CA: Stanford University Press.

Asian Consulting Engineers. 2016. *Environmental Impact Assessment (EIA) Report for Development Drilling of Wells in Onshore ML Areas of Jorhat and Golaghat Districts, Assam State.* New Delhi: Oil and Natural Gas Corporation.

Asoso, Yonuo. 1974. *The Rising Nagas: A Historical and Political Study.* Jaipur: Vivek Publishing House.

Balibar, Étienne. 1994. *Masses, Classes, Ideas: Studies on Politics and Philosophy before and after Marx.* London: Routledge.

Banerjee, Paula, ed. 2008. *Women in Peace Politics.* New Delhi: Sage Publications.

Banerjee, Paula, and Anasua Basu Ray Chaudhury, eds. 2011. *Women in Indian Borderlands.* New Delhi: Sage Publications.

Barbora, Sanjay. 2005. "Autonomy in the Northeast." In *The Politics of Autonomy: Indian Experience,* edited by Ranabir Samaddar. New Delhi: Sage.

Barpujari, H. K. 1986. *The American Missionaries and North-East India,1836– 1900 AD: A Documentary Study.* Guwahati: Spectrum.

Barth, Fredrik. 1969. *Ethnic Groups and Boundaries.* Boston: Little, Brown.

Baruah, Ditee Moni. 2011. "The Refinery Movement in Assam." *Economic and Political Weekly* 46 (1): 63–69.

Baruah, Sanjib. 1999. *India against Itself: Assam and the Politics of Nationality*. Philadelphia: University of Pennsylvania Press.

———. 2001. "Clash of Resource Use Regimes in Colonial Assam: A Nineteenth Century Puzzle Revisited." *Journal of Peasant Studies* 28 (3): 109–24.

———. 2003. "Nationalizing Space: Cosmetic Federalism and the Politics of Development in Northeast India." *Development and Change* 34 (5): 915–39.

———. 2005. *Durable Disorder: Understanding the Politics of Northeast India*. New Delhi: Oxford University Press.

———. 2012. *Beyond Counter-Insurgency: Breaking the Impasse in Northeast India*. New Delhi: Oxford University Press.

Baxi, Pratiksha. 2009. *Habeas Corpus: Judicial Narratives of Sexual Governance*, Working Paper Series. New Delhi: Centre for the Study of Law and Governance, Jawaharlal Nehru University. https://www.jnu.ac.in/sites/default/files/u63/09-Habeus%20%28Pratiksha%20Baxi%29.pdf.

Behal, Rana. 1985. "Forms of Labour Protest in Assam Valley Tea Plantations, 1900–1930." *Economic and Political Weekly* 20 (4): 19–26.

Beteille, Andre. 1986. "The Concept of Tribe with Special Reference to India." *European Journal of Sociology* 27 (2): 296–318.

Bhagabati, Ananda C. 1993. "Social Formations in North-East India." In *People of India*, edited by S. B. Roy and Asok Ghosh. New Delhi: Inter-India.

Borthakur, Brig. R. 2015. "Internal Security Scenario of North East India." *Indian Defence Review* 30 (1).

Bourdieu, Pierre. 1990. *The Logic of Practice*. Stanford, CA: Stanford University Press.

Bruck, Gabeiele vom, and Barbara Bodenhorn, eds. 2006. *An Anthropology of Names and Naming*. Cambridge: Cambridge University Press.

Butalia, Urvashi. 2000. *The Other Side of Silence: Voices from the Partition of India*. Durham, NC: Duke University Press.

Chakrabarti, S. K, H. J. Singh, M. D. S. Akhtar, R. K. Singh, Dominique Silox, Tohon Polanco-Ferrer, and Rosa E. Polanco-Ferrer. 2011. "Structural Style of the Assam Shelf and Schuppen Belt, A &AA Basin, India." Paper presented at the Second South Asian Geoscience Conference and Exhibition, GEO-India, Jan. 12–14, 2011, New Delhi. http://www.searchanddiscovery.com/pdfz/documents/2011/50409chakrabarti/ndx_chakrabarti.pdf.html.

Chakrabarty, Dipesh. 2002. *Habitations of Modernity: Essays in the Wake of Subaltern Studies*. Chicago: Chicago University Press.

Chatterjee, Partha. 1993. *The Nation and Its Fragments: Colonial and Postcolonial Histories*. Princeton, NJ: Princeton University Press.

———. 2004. *The Politics of the Governed: Reflections on Popular Politics in Most of the World*. New York: Columbia University Press.

Chaube, S. K. 2012. *Hill Politics in Northeast India*. 3d ed. New Delhi: Orient Blackswan.

Cohn, Bernard. 1996. *Colonialism and Its Forms of Knowledge: The British in India*. Princeton, NJ: Princeton University Press.

———. 2004. *The Bernard Cohn Omnibus*. New Delhi: Oxford University Press.

Cole, Jennifer, and Lynn Thomas, eds. 2009. *Love in Africa*. Chicago: University of Chicago Press.

Cooper, Frederick. 1993. "Colonizing Time: Work Rhythms and Labor Conflict in Colonial Mombasa." In *Colonialism and Culture*, edited by Nicholas B. Dirks. Princeton, NJ: Princeton University Press.

Coronil, Fernando. 1997. *The Magical State: Nature, Money, and Modernity in Venezuela*. Chicago: University of Chicago Press.

Cronon, William. 1993. "Kennecott Journey: The Path out of Town." In *Under an Open Sky: Rethinking America's Western Past*, edited by William Cronon, George A. Miles, and Jay Gitlin. New York: W. W. Norton.

de Certeau, Michel, Luce Giard, and Pierre Mayol. 1998. *The Practice of Everyday Life: Volume 2, Living and Cooking*. Minneapolis: University of Minnesota Press.

Deka, Harekrishna. 2008. "North-East in Fragments and Invention of a Metaphor." In *Souvenir: North East India History Association (29th Annual Session)*, edited by Chandan Kumar Sarma. Dibrugarh: Department of History, Dibrugarh University.

Directorate General of Hydrocarbons. n.d. *Basin Information: Assam-Arakan Basin*. Noida: Ministry of Petroleum and Natural Gas, Government of India.

Douglas, Mary. 1966. *Purity and Danger: An Analysis of Concepts of Pollution and Taboos*. London: Routledge and Kegan Paul.

Durham, Deborah. 2011. "Disgust and the Anthropological Imagination." *Ethnos* 76 (2): 131–56.

Dutta, Julee. 2016. "Centre-Assam Financial Relations: A Critical Analysis." *IOSR Journal of Humanities and Social Science* 21 (4): 46–51.

Eilenberg, Michael. 2012. *At the Edge of the State: Dynamics of State Formation in the Indonesian Borderlands*. Leiden: KITLV Press.

Escobar, Arturo. 1995. *Encountering Development: The Making and Unmaking of the Third World*. Princeton, NJ: Princeton University Press.

Ferguson, James. 1994. *The Anti-Politics Machine: "Development," Depoliticization, and Bureaucratic Power in Lesotho*. Minneapolis: University of Minnesota Press.

———. 1999. *Expectations of Modernity: Myths and Meanings of Urban Life on the Zambian Copperbelt*. Berkeley: University of California Press.

———. 2006. *Global Shadows: Africa in the Neoliberal World Order*. Durham, NC: Duke University Press.

Ferguson, James, and Akhil Gupta. 2002. "Spatializing States: Towards an Ethnography of Neoliberal Governmentality." *American Ethnologist* 24 (9): 981–1002.

Furer-Haimendorf, Christoph von. 1962. *The Naked Nagas.* Calcutta: Thacker Spink.

———. 1976. *Return to the Naked Nagas: An Anthropologist's View of Nagaland 1936–1970.* London: John Murray.

Geary, Patrick. 1986. "Sacred Commodities: The Circulation of Medieval Relics." In *The Social Life of Things: Commodities in Cultural Perspectives,* edited by Arjun Appadurai. Chicago: Chicago University Press.

Geertz, Clifford. 1980. *Negara: The Theatre State in Nineteenth-Century Bali.* Princeton, NJ: Princeton University Press.

Geertz, Clifford, Hildred Geertz, and Lawrence Rosen. 1979. *Meaning and Order in Moroccan Society: Three Essays in Cultural Analysis.* Cambridge: Cambridge University Press.

Ghosh, Kaushik. 1999. "A Market for Aboriginality: Primitivism and Race Classification in the Indentured Labour Market of Colonial India." In *Subaltern Studies 10: Writings on South Asian History and Society,* edited by Ranajit Guha. New Delhi: Oxford University Press.

Ghurye, S. C. 1963. *The Scheduled Tribe.* Bombay: Popular Press.

Goswami, B. B. 1993. "Dimensions of the Naga Culture and Change." In *People of India,* edited by S. B. Roy and Asok K. Ghosh. New Delhi: Inter-India.

Goswami, Manu. 2004. *Producing India: From Colonial Economy to National Space.* Chicago: Chicago University Press.

Government of Nagaland. 1976. *Report on Evaluation Study on Medium Sized Farm in Merapani.* Kohima: Department of Planning and Evaluation, Government of Nagaland.

———. 2000. "Report on Evaluation Study of Medium Sized Farm in Merapani—Merapani Medium Sized Seed Farm." In *Compendium of Evaluation Studies Conducted by Directorate of Evaluation (1970–2000).* Kohima: Government of Nagaland.

———. 2005. *Nagaland: The State with a Difference—Tourism Brochure.* Kohima: Directorate of Tourism, Nagaland.

———. 2006. *Adding Value to Shifting Cultivation in Nagaland, India.* New Delhi: Nagaland Empowerment of People through Economic Development.

———. 2007. *Evaluation Study "On Employment Opportunities Foregone By Nagas and Employment of Non-Nagas in the State."* Kohima: Directorate of Evaluation, Nagaland.

———. 2009. *District Human Development Report, Kohima: Strengthening of State Plans for Human Development.* Kohima: Department of Planning and Coordination, Nagaland.

———. N.d. *Investment Opportunities in Mineral Sector in Nagaland.* Dimapur: Directorate of Geology and Mining, Government of Nagaland.

Graeber, David. 2001. *Towards an Anthropological Theory of Value: The False Coin of Our Dreams.* New York: Palgrave.

Gudeman, Stephen, and Alberto Rivera. 1990. *Conversations in Columbia: The Domestic Economy in Life and Text.* Cambridge: Cambridge University Press.

Guha, Ranajit. 1997. *A Subaltern Studies Reader, 1986–1995.* Minneapolis: University of Minnesota Press.

Gupta, Akhil. 1995. "Blurred Boundaries: The Discourse of Corruption, the Culture of Politics, and the Imagined State." *American Ethnologist* 22 (2): 375–402.

Gupta, Akhil, and James Ferguson, eds. 1997. *Anthropological Locations: Boundaries and Grounds of a Field Science.* Berkeley: University of California Press.

Hansen, Thomas Blom. 2001. *Wages of Violence: Naming and Identity in Postcolonial Bombay.* Princeton, NJ: Princeton University Press.

———. 2006. "Where Names Fall Short: Names as Performances in Contemporary Urban South Africa." In *The Anthropology on Names and Naming*, edited by Gabriele vom Bruck and Barbara Bodenhorn. Cambridge: Cambridge University Press.

Hansen, Thomas Blom, and Finn Stepputat. 2001. *States of Imagination: Ethnographic Explorations of the Postcolonial State.* Durham, NC: Duke University Press.

———, eds. 2005. *Sovereign Bodies: Citizens, Migrants, and States in the Postcolonial World.* Princeton, NJ: Princeton University Press.

Haraway, Donna. 1988. "Situated Knowledge: The Science Question in Feminism and the Privilege of Partial Perspective." *Feminist Studies* 14 (3): 575–99.

Harris, Olivia, ed. 1996. *Inside and Outside the Law.* New York: Routledge.

Harvey, David. 2000. "Cosmopolitanism and the Banality of Geographical Evils." *Public Culture* 12 (2): 529–64.

———. 2006. *Spaces of Global Capitalism: A Theory of Uneven Geographical Development.* New York: Verso.

Hazarika, Sanjoy. 1994. *Strangers of the Mist: Tales of War and Peace from India's Northeast.* New Delhi: Penguin.

Hefner, Robert. 1993. *The Political Economy of Mountain Java: An Interpretative History.* Berkeley: University of California Press.

Hirschman, Albert. 1977. *The Passions and the Interest.* Princeton, NJ: Princeton University Press.

Ho, Engseng. 2006. *The Graves of Tarim: Genealogy and Mobility across the Indian Ocean.* Berkeley: University of California Press.

Hutton, John Henry. 1921. *The Angami Nagas.* London: Macmillan.

Hyden, Goran. 1983. *No Shortcuts to Progress: African Development in Perspective.* Berkeley: University of California Press.

Iralu, Kaka D. 2001. *Nagaland and India: The Blood and the Tears.* Nagaland: Kaka D. Iralu.

Kar, Bodhisattva. 2009. "When Was the Postcolonial?" In *Beyond Counter-insurgency: Breaking the Impasses in Northeast India*, edited by Sanjib Baruah. New Delhi: Oxford University Press.

Karlsson, Bengt G. 2011. *Unruly Hills: A Political Ecology of India's Northeast*. New Delhi: Orient Blackswan and Social Science Press.

Kathipri, Athili, and Temjenrenla Kechu. 2011. "Infrastructure and Connectivity." In *Kohima District Human Development Report*. Kohima: Department of Planning and Coordination, Government of Nagaland.

Kelly, Tobias. 2008. "The Attractions of Accountancy: Living an Ordinary Life during the Second Palestinian Intifada." *Ethnography* 9 (3): 351–76.

Kikon, Dolly. 2004. *Experiences of Naga Women in Armed Conflict: Narratives from a Militarized Society*. New Delhi: WISCOMP, Foundation for Universal Responsibility of His Holiness the Dalai Lama.

———. 2009. "From Loin Cloths to Olive Greens: The Politics of Clothing the Naked Nagas." In *Beyond Counter-insurgency: Breaking the Impasses in Northeast India*, edited by Sanjib Baruah. New Delhi: Oxford University Press.

Kikon, Dolly, and Duncan McDuie-Ra. 2017. "English-Language Documents and Old Trucks: Creating Infrastructure in Nagaland's Coal Mining Villages." *South Asia: Journal of South Asian Studies* 40 (4): 772–91.

Knox, T. M. 1967. *Hegel's Philosophy of Right*. London: Oxford University Press.

Leach, Edmund. 1954. *Political Systems of Highland Burma: A Study of Kachin Social Structure*. Cambridge, MA: Harvard University Press.

Lefebvre, Henri. 1991. *The Production of Space*. Oxford: Blackwell.

Li, Tania. 2007. *The Will to Improve: Governmentality, Development, and the Practice of Politics*. Durham, NC: Duke University Press.

———. 2014. *Land's End: Capitalist Relations on an Indigenous Frontier*. Durham, NC: Duke University Press.

Lotha, Abraham. 2007. *History of Naga Anthropology (1832–1947)*. Dimapur: Chumpo Museum.

———. 2009. "Articulating Naga Nationalism." PhD diss., City University of New York.

Ludden, David. 1992. "India's Development Regime." In *Colonialism and Culture*, edited by Nicholas Dirks. Ann Arbor: University of Michigan Press.

Lutgendorf, Philip. 2007. *Hanuman's Tale*. New York: Oxford University Press.

Malkki, Liisa. 1992. "National Geographic: The Rooting of Peoples and the Territorialization of National Identity among Scholars and Refugees." *Cultural Anthropology* 7 (1): 24–44.

———. 1995. *Purity and Exile: Violence, Memory, and National Cosmology among Hutu Refugees in Tanzania*. Chicago: Chicago University Press.

Massey, Doreen. 1994. *Space, Place, and Gender*. Minneapolis: University of Minnesota Press.

————. 1995. *Spatial Divisions of Labor: Social Structures and the Geography of Production.* New York: Routledge.

————. 2005. *For Space.* London: Sage Publications.

Mauss, Marcel. 1990. *The Gift: The Form and Reason for Exchange in Archaic Societies.* New York: W. W. Norton.

Mbembe, Achille. 2001. *On the Postcolony.* Berkeley: University of California Press.

————. 2005. "Sovereignty as a Form of Expenditure." In *Sovereign Bodies: Citizens Migrants and States in the Postcolonial World,* edited by Thomas Blom Hansen and Finn Stepputat. Princeton, NJ: Princeton University Press.

McDuie-Ra, Duncan. 2012. *Northeast Migrants in Delhi: Race, Refuge, Retail.* Amsterdam: Amsterdam University Press.

McDuie-Ra, Duncan, and Dolly Kikon. 2016. "Tribal Communities and Coal in Northeast India: The Politics of Imposing and Resisting Mining Bans." *Energy Policy* 99 (C): 261–69.

Ministry of Petroleum and Natural Gas. 2016. *Indian Petroleum and Natural Gas Statistics 2015–16.* New Delhi: Government of India.

Mintz, Sydney. 1985. *Sweetness and Sugar: The Place of Sugar in Modern History.* New York: Penguin.

Mitchell, Timothy. 2002. *Rule of Experts: Egypt, Techno-Politics, Modernity.* Berkeley: University of California Press.

————. 2011. *Carbon Democracy: Political Power in the Age of Oil.* New York: Verso.

Nash, June. 1993. *We Eat the Mines and the Mines Eat Us: Dependency and Exploitation in Bolivian Tin Mines.* Rev. ed. New York: Columbia University Press.

Newman, David. 2006. "The Lines That Continue to Separate Us: Borders in Our 'Borderless' World." *Progress in Human Geography* 30 (2): 143–61.

Nugent, Paul. 2002. *Smugglers, Secessionists, and Loyal Citizens on the Ghana-Togo Frontier.* Athens: Ohio University Press.

Oil and Natural Gas Corporation. 2009. *Assam and ONGC: Synergy of Over 50 Years, Nazira.* New Delhi: Oil and Natural Gas Corporation.

Phukon, Girin, ed. 2003. *Ethnicisation and Politics in Northeast India.* New Delhi: South Asia Publications.

————. 2010. *Documents on Ahom Movement in Assam.* Moranhat: Institute of Tai Studies and Research Publication.

Povinelli, Elizabeth. 2002. *The Cunning of Recognition: Indigenous Alterities and the Making of Australian Multiculturalism.* Durham, NC: Duke University Press.

————. 2006. *The Empire of Love: Toward a Theory of Intimacy, Genealogy, and Carnality.* Durham, NC: Duke University Press.

Prescott, J. R. V. 1987. *Political Frontiers and Boundaries.* London: Allen and Unwin.

Raffles, Hugh. 2002. *In Amazonia: A Natural History.* Princeton, NJ: Princeton University Press.

Rajaram, Prem Kumar, and Carl Grundy-Warr. 2007. *Borderscapes: Hidden Geographies and Politics at Territory's Edge.* Minneapolis: University of Minnesota Press.

Reeves, Madeleine. 2005. "Locating Danger: Konfliktologiia and the Search for Fixity in the Ferghana Valley Borderlands." *Central Asian Survey* 24 (1): 67–81.

Riles, Annelise. 2006. *Documents: Artifacts of Modern Knowledge.* Ann Arbor: University of Michigan Press.

Rousseau, Jean-Jacques. 1984. *Of the Social Contract or Principles of Political Right & Discourse on Political Economy.* New York: Harper and Row.

Roy-Burman, B. K. 1983. "Transformation of Tribes and Analogous Social Formation." *Economic and Political Weekly* 18 (27): 1172–74.

Saikia, Arupjyoti. 2011. "Kaziranga National Park: History, Landscape and Conversation Practices." *Economic and Political Weekly* 46 (32): 12–13.

Saikia, Yasmin. 2004. *Fragmented Memories: Struggling to Be Tai Ahom in India.* Durham, NC: Duke University Press.

———. 2011. *Women, War, and the Making of Bangladesh.* Durham, NC: Duke University Press.

Samaddar, Ranabir, ed. 2001. *A Biography of the Indian Nation, 1947–1997.* New Delhi: Sage.

———. 2005. *The Politics of Autonomy: Indian Experience.* New Delhi: Sage.

Sanyu, Visier. 1996. *A History of Nagas and Nagaland: Dynamics of Oral Tradition in Village Formation.* New Delhi: Commonwealth.

Sawyer, Suzana. 2004. *Crude Chronicles: Indigenous Politics, Multinational Oil, and Neoliberalism.* Durham, NC: Duke University Press.

Scott, James C. 1998. *Seeing Like a State.* New Haven, CT: Yale University Press.

———. 2009. *The Art of Not Being Governed: An Anarchist History of Upland Southeast Asia.* New Haven, CT: Yale University Press.

Shah, Alpa. 2010. *In the Shadows of the State: Indigenous Politics, Environmentalism, and Insurgency in Jharkhand, India.* Durham, NC: Duke University Press.

Sharma, Jayeeta. 2009. "'Lazy' Natives, Coolie Labour, and the Assam Tea Industry." *Modern Asian Studies* 43 (6): 1287–1324.

———. 2011. *Empire's Gardens: Assam and the Making of India.* Durham, NC: Duke University Press.

Siegel, James. 2011. *Objects and Objections of Ethnography.* New York: Fordham University Press.

Simmel, Georg. 1971. *On Individuality and Social Forms: Selected Writings.* Chicago: University of Chicago Press.

Sivaramakrishnan, K. C. 1999. *Modern Forests: State Making and Environmental Change in Eastern India*. Stanford, CA: Stanford University Press.

Sivaramakrishnan, K. C., and Arun Agrawal. 1997. *Regional Modernities: The Cultural Politics of Development in India*. Stanford, CA: Stanford University Press.

Skaria, Ajay. 1997. "Shades of Wildness: Tribe, Caste, and Gender in Western India." *Journal of Asian Studies* 56 (3): 726–45.

———. 1999. *Hybrid Histories: Forests, Frontiers, and Wildness in Western India*. New York: Oxford University Press.

Spivak, Gayatri Chakravorty. 1999. *A Critique of Postcolonial Reason*. Cambridge, MA: Harvard University Press.

Stepputat, Finn. 2004. "Marching for Progress: Rituals of Citizenship, State, and Belonging in a High Andes District." *Bulletin of Latin American Research* 24 (2): 244–59.

Stoler, Ann. 2002. *Carnal Knowledge and Imperial Power: Race and the Intimate in Colonial Rule*. Berkeley: University of California Press.

Sturgeon, Janet C. 2005. *Border Landscapes: The Politics of Akha Land Use in China and Thailand*. Seattle: University of Washington Press.

Tambiah, Stanley J. 1977. "The Galactic Polity: The Structure of Traditional Kingdoms in Southeast Asia." *Annals of the New York Academy of Sciences* 293 (1): 69–97.

Taussig, Michael. 1980. *The Devil and Commodity Fetishism in South America*. Chapel Hill: University of North Carolina Press.

———. 1997. *The Magic of the State*. New York: Routledge.

———. 2006. *Walter Benjamin's Grave*. Chicago: Chicago University Press.

Thiranagama, Sharika. 2011. *In My Mother's House: Civil War in Sri Lanka*. Philadelphia: University of Pennsylvania Press.

Thiranagama, Sharika, and Tobias Kelly, eds. 2010. *Traitors: Suspicion, Intimacy, and the Ethics of State-Building*. Philadelphia: University of Pennsylvania Press.

Thompson, E. P. 1963. *The Making of the English Working Class*. New York: Vintage Books.

———. 1967. "Time, Work Discipline, and Industrial Capitalism." *Past and Present* 38: 56–97.

———. 1975. *Whigs and Hunters: The Origins of the Black Act*. New York: Pantheon Books.

Tsing, Anna Lowenhaupt. 1993. *In the Realm of the Diamond Queen: Marginality in an Out-of-the-Way Place*. Princeton, NJ: Princeton University Press.

———. 2003. "Inside the Economy of Appearances." *Public Culture* 12 (1): 115–44.

———. 2005. *Friction: An Ethnography of Global Connections.* Princeton, NJ: Princeton University Press.

van Schendel, Willem. 2002. "Geographies of Knowing, Geographies of Ignorance: Jumping Scale in Southeast Asia." *Environment and Planning D: Society and Space* 20: 647–68.

———. 2005. *The Bengal Borderland: Beyond State and Nation in South Asia.* London: Anthem.

van Schendel, Willem, and Itty Abraham, eds. 2005. *Illicit Flows and Criminal Things: State, Border, and the Other Side of Globalization.* Bloomington: Indiana University Press.

Verghese, B. G. 1996. *India's Northeast Resurgent: Ethnicity, Insurgency, Governance, Development.* New Delhi: Konark.

von Stockhausen, Alban. 2015. "Watlong, the Naga Queen: Negotiating Local Identities through Narratives." *Asian Ethnicity* 17 (3): 353–69.

Watts, Michael J. 2015. "Spaces of Insurgency: Power, Place and Spectacle in Nigeria." In *Geographies of Power: Recognizing the Present Moment of Danger,* edited by Heather Merrill and Lisa Hoffman. Athens: University of Georgia Press.

Xaxa, Virginius. 1999. "Tribes as Indigenous People of India." *Economic and Political Weekly* 34 (51): 3589–95.

Yanagisako, Sylvia, and Carol Delaney. 1994. *Naturalizing Power: Essays in Feminist Cultural Analysis.* New York: Routledge.

Yeats, W. B. 1989. *The Collected Works of W. B. Yeats, Volume 1: The Poems Revised.* New York: Macmillan.

Zamindar, Vazira Fazila-Yacoobali. 2007. *The Long Partition and the Making of Modern South Asia: Refugees, Boundaries, Histories.* New York: Columbia University Press.

INDEX

Adivasi: as *bagania*, 55; as an ethnic category, 42; Gorejan village established by, 39; indigenous people of Hindi and Bengali as, 159n2; and Roman Catholicism, 39–40; and scheduled tribe status, 118, 166n4; as strangers in Naga villages, 38, 118; and tea plantation labor, 55–56, 159n2. *See also* Gorejan village

Adivasi—individual stories of: Birsa Munda, 117–19; Gorib Maji, 116, 117*fig.*, 119; inter-ethnic friendship on Aka's *jhum* fields, 111–15, 113*fig.*; Samar, 39–43

affection: everyday practices of, 62. *See also* friends and friendship; *morom*; state love

Agrawal, Arun, 157–58n3

agriculture and land issues: and Nagaland Seed Farm, 82–85; and oil seepage and spills left by abandoned ONGC oil wells, 96, 111, 124–25, 167n5; state subsidies in Assam, 65; subsistence agriculture, 111, 125, 129. *See also howri* (labor exchange); *jhum* fields; tea plantations; wet rice cultivation

Ahmed, Romizuddin, 98, 99

Ahmed, Sara, 160n2

Ahom dynasty, 158–59n11; attack by Maan-Singhpo, 15, 158n10; Dalimi's absence from the *Buranji* chronicles, 54; Jaymati's absence from the *Buranji* chronicles, 54; and the Phoms, 15. *See also* Dalimi and Gadapani

Ahoms, dietary practices of, 97–98

Ali, Molong (coal trader in Sonari town): friendship with Wangcho, 99–101; love for his wife Begum, 59–61

All Assam Students Union (AASU), 141

Amnesty International, 26

Anaki, 4*map*, 6*map*, 90, 137

Anaki Yimsen, 3, 126, 127

Appadurai, Arjun, 81, 107

Armed Forces Special Powers Act (AFSPA), 13, 17–18, 72, 73, 75, 136, 150

Article 371A of the Indian Constitution, 5, 22, 71–73, 79, 104, 110, 118, 121, 164–65n14

Arunachal Pradesh, 4*map*, 23, 28, 67, 163n2–3, 165n20

Assam: colonial administration of, 69; dietary history and tastes of, 96–98; disappearance of trader Atul Rajkhowa, 92–95; Dispur

179

Assam (*continued*)
(capital), 64, 67; drought management system in, 64; and Naga villagers, 118; *Oxom-baxi* (inhabitants of Assam), 67; tea plantations in, 6*map*, 41. *See also* Adivasi; Assam oil fields; Assam Police; Brahmaputra Valley; foothill border between Assam and Nagaland; Guwahati; Singibil village; Sonari (Assam)—stories of residents of; Upper Assam (Ujoni Assam)

Assamese: affluence of, 95; and intermarriage, 50; origin myth about Naga and Assamese brothers, 14–15; taste preferences of, 96–97. *See also* Ali, Molong (coal trader in Sonari town); Sonari (Assam)—stories of residents of

Assam oil fields, 6*map*; and the Assam-Arakan Basin, 23; contentious politics associated with, 10–13, 135–37; and international bidders, 13; and the Oil and Natural Gas Commission Act of 1959, 11; petroleum discovered around 1825, 11; and the Refinery Movement (1956–57), 12; and regional identity, 11–13, 21; Upper Assam oil and gas fields, 24, 137

Assam Police, 3, 18–19; checkpoint at Singibil *haat*, 92–94

Balibar, Étienne, 50
Baruah, Ditee Moni, 11
Baruah, Sanjib, 69, 165n20
Baxi, Pratiksha, 161n5
Behal, Rana, 39, 41–42, 159n4
Bezbarua, Lakshminath, 54
Bhandari, 4*map*, 6*map*; Naga cattle brokers described by Chumbemo, 90–91

Bihari, 9, 134, 151; association of Bihari traders with the underground, 148

boundaries and borders: ambiguous continuity of, 8, 28–29, 102; and nation-states, ix–x; and power structures, 157n1; and territorial boundaries drawn on ethnic bodies, 47–48. *See also* foothill border between Assam and Nagaland; frontiers; sovereignty and power structures; strangers

Brahmaputra Valley, 16, 29, 36, 91, 95, 96; enclave economy of, 5, 21, 47; and the geology of the Assam-Arakan Basin, 23–24; oil discovered in Makum, 21; and state love, 64; and the story of Dalimi and Gadapani, 51, 53–55; tea production in, 41

British: administration of villages, 29–32; coal mining operations in Naginimora, 52–53, 77; and the Inner Line Regulation (1873), 5. *See also* colonialism

Burma: Edmund Leach's work on, 15–16, 75; Maan as a name for the Burmese, 158–59n11

carbon citizenship: "good, reliable contractors" identified by the ONGC, 142–43; testing of, 143–44, 149–50

Central Industrial Security Forces (CISF), 3, 13, 18, 118; and the murder of Nilikesh Gogoi and Bulu Gogoi, 26–27; tea, oil, and gas viewed as neutral commodities, 75–76

Central Reserve Police Force (CRPF): Disputed Area Belt controlled by, 18, 74; foothill border between Assam and Nagaland monitored

by, 3, 13, 18, 74, 75, 118; and the
Nagaland Seed Farm, 82
Champang ONGC: carbon fantasies
of Yan (resident of), 121–22, 154;
oil seepage and spills left by abandoned ONGC oil wells, 39, 96, 111,
121. *See also* Kithan, Mr. (Champang ONGC resident)
checkpoints: in Champang ONGC,
123; and harvest time, 102; photo-
graphing attempted by author at,
20; at Singibil *haat*, 92–94; and
tax collection, 104; ubiquity of, 3,
17, 18–19, 28, 66, 73, 74–75, 143–44,
154. *See also* militarized sociality;
surveillance
Chi village: chief minister's (CM)
state visit to, 77–79, 77*fig*.; as a
symbol of state love, 76–79
Christianity: and Adivasi, 39–40,
159–60n5; biblical relationship
between suffering and spirituality,
56–58; church in Longtssori village, 34–35, 37
citizenship: of Marwari traders, 147–
49; and security profiling, 144–47.
See also carbon citizenship
cleanliness and hygiene: of Assamese
compared with Nagas, 95; of
Nagas described by Molong Ali,
99–101. *See also* ethnic purity
coal mines, 6*map*, 133*fig*.; artisanal,
126; British coal mining operations in Naginimora, 52–53, 77;
and the Coal Act of 2006, 131; and
coal taxes, 131–32; "hypermasculine militarism" of coal extraction, 128, 130–31; and speculation,
128, 134; and territorial boundaries drawn on ethnic bodies, 47–48
Cole, Jennifer, 162n10
colonialism: and the administration
of Assam, 69; and English used in

official documents, 105–6; Indian
state's nationalizing of frontiers,
165n20; and the Inner Line Regulation (1873), 5; and the Marwaris,
168n7; oil exploration as an example of Indian internal colonialism,
10, 21
Cronon, William, 43
culture: as a locus for naturalizing
power, practices, and hierarchies,
51, 162n7; of the Nagas preserved
by the Konyak Nagas, 76

Dalimi and Gadapani: Dalimi identified as Watlong, 55; and the political life of Naginimora, 53–55,
162n9; story of, 51–52
dhorom (Christian faith), 57, 162n11
Dikhow river, 53, 55
Dimapur, 149; Directorate of Geology
and Mining in, 131–32
Disputed Area Belt, 18
Disturbed Area Act (1955), 13, 73, 75,
136, 150
documents: and British administration of villages, 29–32; documentary performances, 105–7;
verification at checkpoints of, 20.
See also gazettes; Indian Constitution; Sixteen-Point Agreement
domestic abuse: and concepts of ethnic purity, 48–50; and *morom*,
56–58, 62
Douglas, Mary, 160–61n2
Dutta, Julee, 11

education. *See* schools and education
English language, and official documents, 105–6
environmental issues: degradation
due to resource extraction, 76–
77, 166n1, 167n5; and the development of frontier regions, 165n20;

environmental issues (*continued*)
oil seepage and spills left by
abandoned oil wells, 18*fig.*, 39,
96, 111, 121; and ONGC seismic
operations in Sibsagar district,
142; and the power, relations,
and politics of environmentality,
9, 157–58n3

ethnic groups and alliances: hill
states viewed as primarily home
to one ethnic group, 163n3. *See
also* Ahom; Bihari; Naga; Nepalis;
United Liberation Front of Assam

ethnic purity: and camp refugees, 50,
161–62n6; and the case of Lulu's
intermarriage, 49–50, 55, 62; and
male and female bodies, 48; and
the story of Dalimi and Gadapani,
51–55

family members (*ghor manu*) and
family ties: adoptions of *ghor
manu nisena* (just like a household member), 102, 107–8, 111–12,
116–18, 130, 154; and Naga cultural
values, 129–31; and trading networks, 107–8. *See also* domestic
abuse

Ferguson, James, 139

foothill border between Assam and
Nagaland, 4*map*; Assam-Arakan
Basin as the geological name for,
7; everyday dangers of life in,
25–27, 92–93, 102–3, 117–18; and
the Inner Line Regulation (1873),
5; monitoring by the Central
Reserve Police Force (CRPF), 3,
13, 74, 75–118; production of spaces
of extraction in, 8; schools in, 151–
54, 152*fig.*, 153*fig.*; and state surveillance activities, 3, 17–20, 28–
29, 66, 72–76, 84, 110, 136–37, 150,
154, 166n3; tensions between the
Assamese and the Nagas, 35–37,
82–84, 92–95

friends and friendship: and *haats*, 19,
98–101; inter-ethnic friendship on
Aka's *jhum* fields, 111–15, 113*fig.*;
Molong Ali and Wangcho, 99–101;
seasonal, 103–5

frontiers: abundant resources associated with, 10, 123–24; and colonial administration of Assam,
69; conceptions of, ix; as global
hotspots of biodiversity, 165n20;
as a place of capital, extraction,
and violence, 9; production of,
157n2. *See also* boundaries and
borders

gazettes: and colonial administration, 31, 159n1; declarations of
"disturbed" areas announced in,
17; *Nagaland Gazette* on the Nagaland Coal Policy (Amendment) of
2014, 131; recognition of new villages announced in, 34, 126

Gelakey town, 4*map*, 6*map*, 19,
25–27, 102, 105, 137

gender: and coal trading, 129–31;
exclusion of Naga women from
positions of power and decision-
making forums, 128–29; and
notions of ethnic purity, 48, 49.
See also masculinity; patriarchy/
patriarchal

geology: of the Assam-Arakan Basin,
7–8, 21, 23–24; fault lines in
Northeast India, 137–38; and
the ONGC oil carnival, 136*fig.*,
137–39

Ghosh, Kaushik, 168n7

Gogoi, Akhil, 12

Gogoi, Nilikesh, 26–27

Gogoi, Rajan, 144–48

Gorejan village, 39–40, 43

Gudeman, Stephen, and Alberto Rivera, 87–88
Guwahati, 35, 67, 149

*haat*s (weekly markets): friendship and connection invoked by, 98–101; *haatkhawa* (tax collectors), 32, 86, 94–95; as historical and cultural signposts, 19; medicine stall of Romizuddin Ahmed, 98, 99; Nagabat *haat* traders, 29, 89–92, 91*fig.*, 97–98; Naginimora *haat*, 87*fig.*; Namsa *haat*, 116; origins of, 87; Singibil *haat*, 92–95; as sites of conflict, 19–20, 95; as zones of social interaction, 19, 86–89, 101. *See also* Rajabari *haat*
Hansen, Thomas Blom, and Finn Stepputat, 66
Himalayas, eastern: geology of the Assam-Arakan Basin, 21, 23–24; production of spaces of extraction in, 8. *See also* foothill border between Assam and Nagaland
howri (labor exchange), 119, 129; and inter-ethnic friendship on Aka's *jhum* fields, 111–15, 113*fig.*
Hyden, Goran, 108

indentured laborers, *coolies*, 41, 160n6
Indian Constitution: Article 21, 17; Article 371A, 5, 22, 71–73, 79, 104, 110, 121, 164–65n14; Sixth Schedule of, 163n4
Indian security forces: and competing sovereign powers in the foothills, 9, 13, 63, 71, 73–76, 85; detaining of Rajan Gogi, 144–48; and hydrocarbon activities, 20, 84, 121, 124, 142, 144, 154; perceptions of trading and business communities, 148–49; residents'

accounts of Assam and Nagaland contrasted with views of, 44, 74. *See also* Central Industrial Security Forces (CISF); checkpoints; surveillance
Indo-Naga conflict: cease fire agreement (1963), 71; cease fire agreement (1997), 71, 73, 121; and love of the Naga homeland, 160–61n3
inequalities: Assamese affluence compared with the poverty of Nagas, 91, 95; and the development of Assam, 12; and gender, 128–29; and land ownership, 116–19; and the origin myth about Naga and Assamese brothers, 14–15, 16; and state love, x, 63–65, 80–82; and the technical and knowledge infrastructure, 12
infrastructure: absence of, 64, 81–82; benefits of the technical and knowledge infrastructure, 12; and environmental issues, 165n20; and oil exploration operations, 39
insurgency: and conflict between Naga armed groups and the government of India, 46, 70–71; and the history of resource extraction and militarization, 13–14, 52, 120; profiling of bodies by police, 147
insurgent groups. *See* Naga insurgents; Naga Socialist Council of Nagaland (NSCN); United Liberation Front of Assam (ULFA)

Jaymati, Queen, 54
jhum fields, 7, 28, 128, 129; and corruption related to government relief in Nagaland, 80; inter-ethnic friendship on Aka's *jhum* fields, 111–15, 113*fig.*; and oil

morom (love): definition of, 45; and experiences with sovereignty and power structures, 63, 85; and the gendered politics of women's bodies, 45–46, 49, 51, 56, 61; invocation of *I love you* compared with, 58–61; story of Molong Ali and Begum, 59–61; violence, oppression, and loss as expressions of, 56–58, 62. *See also* state love

Nagabat, 4*map*, 6*map*; *haat* traders, 29, 89–92, 91*fig.*, 97–98

Naga ethnicity: origin myth about Naga and Assamese brothers, 14–15; poverty of, 95; and punishment from speaking Nagamese and Lotha, 151–54, 153*fig.*; and taste preferences, 96–97. *See also* Kithan, Mr. (Champang ONGC resident)

Naga Hills: discovery of oil in, 21; and the Inner Line Regulation (1873), 5; and the Sixteen-Point Agreement, 163–64n10

Naga insurgents: negotiations with state officials on behalf of the villages, 104–5; restrictions on oil exploration by, 73. *See also* Naga Socialist Council of Nagaland (NSCN); United Liberation Front of Assam (ULFA)

Nagaland, 6*map*; creation of, 71, 79, 164n13; laws governing ownership and use of land in, 22, 47, 72–73, 104, 118, 121, 126, 164–65n14; Longtssori village, 33–38; Longwa village in, 145–46; tastes of produce from Nagaland and Assam, 96; and well subsidies, 63–64, 162n1. *See also* Bhandari; foothill border between Assam and Nagaland; Kohima; Lotha (Naga) villages

Nagaland Armed Police, 3, 18–19, 20

Nagaland Coal Policy Act (2006): and coal prospecting licences, 167n6; Nagaland Coal Policy (Amendment) of 2014, 131

Nagaland Gazette, 131

Nagaland Seed Farm, 82–85

Naga museum, 78–79

Naga National Council (NNC), 71, 164n12

Naga Schuppen belt, 24, 28

Naga Socialist Council of Nagaland (NSCN): factions of, 47, 71, 164n12; and security forces' targeting of local residents, 145, 148; and state love, 64

Naginimora, 4*map*, 6*map*; coal exploration by the Indian state, 120; coal mining in, 52–53, 77; *haat* in, 87*fig.*; and the legend of Dalimi and Gadapani, 51–53; tea plantations, 53

Namsa *haat*, 116

National Rural Employment Guarantee Act of 2005 (NREGA), 80–81

natural resources: frontiers associated with, 10, 123–24; and self-determination rights, 21; tea as, 12, 53, 75–76. *See also* Assam oil fields; coal mines

Nazira, 25, 139–41

Nepalis: as indigenous to Nagaland, 118, 148; livelihood struggle of, 9

networks, alliances, and rivalries: and the Ahom kingdom, 158–59n11; and hospitality, 105, 108, 115; and membership in diverse cultural and political organizations, 40; and resource extraction, 7–8, 104, 108, 110–11. *See also* ethnic groups and alliances; *howri* (labor exchange); reciprocity; strangers and others (*oyem*); trading networks; United Liberation Front of Assam

CULTURE, PLACE, AND NATURE
Studies in Anthropology and Environment

www.ingramcontent.com/pod-product-compliance
Lightning Source LLC
Chambersburg PA
CBHW031134270326
41929CB00011B/1624